WTF JUST HAPPENED?!

A sciencey-skeptic explores grief,
healing, and evidence of an afterlife.

WTF Just Happened?!
Book 1

ELIZABETH ENTIN

WTPH Just Happened, LLC

Contents

Author Note	xi
Foreword	xiii
Bob Ginsberg of Forever Family Foundation	
Introduction - WTF JUST HAPPENED?!	xvii
1. How Hard Can Time Travel Really Be?	1
2. I'm Not Done Talking About Reincarnation	10
3. A Class With Someone Who Believes All This Afterlife Stuff	16
4. A Few More Books Get Me Thinking This Isn't All Batshit	21
5. The Sacred Scriptures Of Gary And Julie	24
6. I Go To My First Medium And Wtf?!	36
7. Did I Get A Sign?	48
8. Searching For Psychics And Ghosts	53
9. Weirdness At A Medium Workshop	71
10. I Finally Figure Out What I Am Just Not Seeing	87
11. Spoons, Psychics And WTF?!	94
12. Five-Dollar Readings And Three-Hundred Dollar Candles	111
13. Manifesting and Channeling	121
14. Energy = Mindblown Squared	131
15. Hiding Evidence And Meeting Mediums	141
16. Ordinary People, Extraordinary Claims	151
17. Good And Evil	157
18. The #1 WTF To Top All WTF's That Ever Were	171
19. I'm Not Okay. Okay?	178
20. More Mediumship And Past Life Regressions	193
21. The #2 WTF To Top All Wtf's That Ever Were	210
22. I'm Gonna Keep Trying This Out Of Body Thing	218
23. Just Between Me And My Dad	223
24. Grief Retreat: Not So Weird Anymore	227
25. How In The Actual Fuck??	238
26. Smoke and Mirrors and Séances	242
27. Told-You-So's and Secrets Revealed	272

28. When You Are Going Through Hard Times Just Keep Going	290
29. The Afterlife Luncheon	298
30. The Right Seat on the Right Flight	305
31. In Spite of Stephen Hawking	310
32. Coin Collections and Connections	314
33. The Best Medicine	317
34. Still Getting Enlightened on the Other Side	320
35. It's Not an Addiction. It's Science.	324
Afterword: WTF Do I Even Think?!	327
Teaser: WTF Just Happened?! Book 2	335
Science + Spirituality Salons	339
Stay in Touch!!	341
Notes	343
Also by Elizabeth Entin	347
Acknowledgments	349
About the Author	353

Praise for WTF Just Happened?! Book 1

"Elizabeth Entin embarked on a completely unforeseen journey, after the passing of her beloved father, into the world of mediumship and the scientists who love them. Her memoir is equal parts informative, heart-wrenching, and funny as hell. This soon to be quintessential book on the subject is the perfect guide for anyone who is searching for answers or who has ever had an experience that left them wondering: What the f*ck just happened?!"

> Joe Perreta, Forever Family Foundation Certified Medium (www.joeperreta.com + GOOP)

"The passing of Liz Entin's father sent her on an extraordinary journey.

Devastated by the loss, this scientifically-minded woman, who'd come from a family that was not religious, and did not believe in life after death, suddenly found herself wondering if there might be any science that backed the possibility that the relationship with her beloved father was not forever lost.

After examining the literature around the possibility of time travel-something that might not bring her dad back from the dead, but would at least provide the opportunity to hug him again—and finding that option a less than practical one, she began to look at research that suggested consciousness might not be confined to the physical body, and might even survive its demise.

In this deeply personal recounting of her search for answers, you'll be moved by the love that sent her down this road, and enthralled with what she finds there."

Mike Anthony, Author of Love, Dad: How My Father Died . . . Then Told Me He Didn't and Life at Hamilton. Featured in Netflix "Surviving Death."

"Enthralling! A true page-turner that tells the story of personal transformation backed by science. Liz writes in a way that mimics a relatable internal monologue—so real, questioning, unfiltered, self- aware, and deep. Deep and trying to be open. I thoroughly enjoyed being taken on Liz's journey of discovery for her sake, but also for many others. If you're curious about what comes next or are at any point in your journey of grief after losing a loved one, you'll want to read WTF Just Happened?! "

Stephanie Thoma, Author of Confident Introvert

"As a lay researcher of consciousness, the mind/brain relationship and the possibilities of life after physical death, and as someone who's had the chance to discuss the subject with many of the world's leading academics in the subjects, I try to approach every question with an open but scientifically oriented, empirical mindset. Liz and myself are very similar in this regard, and I have been very impressed with her research since I first came to know of her. Both her podcast and written works give a powerful and personal insight into her research, and her empirical, anti-woo approach has provided both herself and her peers with very strong evidence that our lives do not end at the death of our bodies."

Darren McEnaney, SEEKING I

"Liz Entin's book is captivating and refreshing. Having experience as a Hospice nurse and losing both parents at a young age, this book is highly recommended for any person out there questioning the after life."

Lisa Glaessner, Hospice Nurse

Copyright

Published by WTPH Just Happened, LLC

Copyright © 2021 by Elizabeth H. Entin

Listen to the podcast and learn more about this crazy body of afterlife evidence at www.wtfjusthappened.net

Cover design by Amanda Guerassio of Studio Guerassio

All rights reserved, including the right of reproduction in whole or in part in any form.

This book is based on the author's recollection of true events that transpired. Some names and identifying factors have been changed to protect the privacy of those depicted.

Dedication

Dad, *I had so much fun writing this together. I am sure you were just as shocked as me to discover you were still you. And even more shocked as you watched me figure it out. Thanks for teaching me it's okay to always follow my own path, although you might have drawn the line at this pursuit, while in this dimension.*

Phran, *thank you so much for mentoring me in this crazy world that I never could believe was actually true. Thanks for your patience with my endless questions and listening to my newest theory about how this might all work. I can't wait to see you again when you will (probably) greet me with a big "I told you so" and then make sure I get right to work on the auxiliary board. Love and Misses.*

Author Note

I hope you will enjoy reading the craziest story of my life. The people are real. The events really happened, something I have to frequently remind myself. If you find the events and all I discovered in this book shocking or hard to believe, I am with you!

All mediums, teachers and researchers in this book are themselves.

Identifying details of the guests at the Forever Family Foundation events and other events have been disguised out of respect for their privacy.

The names of my family members (including my dad) and certain identifying details such as the "signs" and "hobbies" have been changed. Details such as my birthday, address, family members' birthdays have also been changed. This is to assure I can still get evidential medium readings and continue my research.

My cousin is a composite character of a few close friends and family members (including actual cousins).

To hear my conversations and me living my best life of asking endless questions of these afterlife experts, (and hopefully getting many of your questions answered) check out my podcast *WTF Just*

Author Note

Happened?!: All about the afterlife. No woo. available on all podcast apps with videos (of most episodes) on YouTube.

www.wtfjusthappened.net

Foreword

Bob Ginsberg of Forever Family Foundation

As is so often the case, the trauma of losing a loved one can result in a search for meaning and, in that pursuit, exploration of things they previously never fathomed. The two questions that we hear the most at Forever Family Foundation are "Does my loved one still exist in some form?" and "Are they OK?"

People have all sorts of beliefs when it comes to the subject of life after bodily death. Some will definitively state that death is final, to think otherwise is purely wishful thinking, and anyone with a modicum of intelligence should realize and accept this fact. I don't judge such people, as I was one of them. However, most people will tell you that they believe in an afterlife. After speaking to such people for the better part of the last twenty years, I have come to realize that many people in this group really mean that they *hope* there is an afterlife. They may have been taught this by their family, their clergy, their cultural or societal influences, but have never really seen any evidence to support their belief. That is fine for some people, but for others like this author, they are unfulfilled and unconvinced. Blind faith does not cut it for them, and uncovering true evidence becomes their mission. Science becomes an obsession and the key not only to acceptance, but the source of hope and comfort.

For many years after my own daughter died, I could not accept the spiritual explanations and personal anecdotes that people were telling me. Of course, with all my heart I wanted to believe that she still existed, as the weight of not having her was crushing the life out of my heart. Fortunately, although a skeptic, I was open-minded. That is quite different than being a closed-minded skeptic, one who will ignore and not accept any evidence, despite what the data reveals. After all these years I remain shocked over the number of scientists who are firmly entrenched in this form of *scientism*. They believe that only science, their view of materialist science, is valid for discovering truth about the nature of reality.

When Liz first came to us in her quest for knowledge, my wife and I of course recognized the depths of her grief, but I saw something in her that resonated with my own journey. Her inquisitive mind in the search for answers was something I understood, and it was unusual for someone at her youthful age. Sure, she was fascinated by mediums and the way in which they worked, but to her credit she realized that mediumship was just one form of evidence suggesting that we are more than our physical bodies. Her studies began in an orderly fashion, and one that I would recommend. To believe in survival of consciousness, you first must come to realize that our minds (consciousness or soul if you prefer) can act independently of our brains. That is a hard concept for many to grasp, but if brain and mind were the same, death would truly be final. Such evidence comes in the form of extra-sensory perception, telepathy, remote viewing, psychokinesis (mind over matter), and a host of other phenomena that defy materialist thinking.

Liz then moved on to study near death experiences, where she learned that people who meet every medical definition of death (no brainwaves, no heartbeat, no respiration, no reflexes) get resuscitated and report clear and lucid thinking, plus all sorts of experiences that could not be realistically explained by medical science. It showed her that something leaves the body at the time of death. She learned about end-of-life experiences and deathbed visions, where

Foreword

people in that window of time just before death see deceased loved ones materialize before them, escorts to the next world. Liz discovered the evidence for reincarnation as she read about studies of children who report past life memories, surprised to learn that these studies have been going on for the past fifty years. She learned about electronic voice phenomena, where discarnates can somehow record voices onto recording devices.

However, as you will read, her main topic of study is with mediums. In full disclosure, even though Liz insisted that all her medium sessions were in the interest of research, more than once my wife Phran chastised her out of fear that Liz was becoming addicted to mediums. Addiction to mediums is rare, but occurs, and includes the common elements of other addictions such as anxiety in between readings and insatiable cravings for connections. I understand it, as those in deep grief can find themselves needing a fix, continuous confirmations that they remain connected to their deceased person. However, the goal, and something that this author has come to realize, is that our loved ones will find a myriad of ways to let us know that they exist, often directly as opposed to through a third person.

The journey that you are about to read represents one person's search for evidence of an afterlife. This is what she needed, and she had the fortitude and intelligence to see it through. In fact, I suspect that her journey has just begun. She enthusiastically tells us what she has learned to date, and you will more than likely smile at her sheer exuberance.

Most importantly, the real significance of this book are not the facts and evidence that she has learned. The take home message is that science and discovery can have significant positive effects on one's grief. As Liz has discovered, science can open the door to recognition and acceptance of things that are non-physical in nature. Science can unlock spirituality. So many of us think that the two terms are mutually exclusive, but have you noticed that some of today's great physicists are starting to sound more like spiritualists? They have learned that we are all connected through waves of information.

Today, Liz is in a far better place than from where she started in her grief, and that growth is a direct result of her changing the way she thinks about death, specifically that it is not final, but a doorway to a continuum of life.

- Bob Ginsberg
Co-founder of Forever Family Foundation
Author of The Medium Explosion and My Life: Here and There

Introduction - WTF JUST HAPPENED?!

I lay back on the couch, took a deep breath, and tried to feel the dead person the psychic medium was sending me. The medium leaned back in a large black leather chair, concentrating on a picture on his phone.

>**Medium:** I am sending you their energy. Let me know what you are getting.

I tried to see if I could feel the "energy" and personality of this dead person. Could I get them to come to me and "tell" me information? I started to feel a soft and gentle energy around me. I felt calm and peaceful. The way you feel when you sit down and grab a glass of wine with a close friend at the end of a stressful day.

>**Medium:** Is it a man or a woman?
>**Me:** Man.
>**Medium:** Anything else…

I breathed deeper, trying to determine what these waves of energy felt like. Lots of love. And warmth. And gentleness.

Introduction - WTF JUST HAPPENED?!

Me: This is a kind person. An especially caring man. Older. Very loving.

I opened my eyes and looked at the medium, expectantly.

Me: So. How did I do?
Medium: Umm, well, actually, it was Hitler.
Me: I told you I didn't have abilities.

I knew I didn't actually have these "defy the laws of physics" abilities. But one time it seemed like I might, which sent me, the biggest skeptic ever, even further down this rabbit hole of afterlife research, the paranormal, or what I often call, really weird shit.

If you had told me a few years ago that I would be befriending mediums, sitting for readings from them, bending spoons, and trying to see if we can connect with dead people, I would never have believed you. Of course, everyone knows mediums are frauds or delusional themselves. And I never had much interest in the woo.

Yet here I was.

The few people who knew about my secret double life exploring evidence of an afterlife, or survival of consciousness, kept telling me that I should write about my experiences.

"But what about the weirdness?" I asked. I mean did I really want my name attached to something so... woo-woo?

"Meh. You'll get over it," my main mentor in this world, Phran, a logical no-nonsense woman, told me when I expressed my worries about other people's opinions if I were to "go public."

Another medium friend Dusten Lyvers agreed. After he accurately described a dresser in my parent's bedroom—which he had never seen before!—he told me, "Look, if I can come out twice, first as gay and then as a medium, you can come out once."

So, I decided to "come out."

But what did I have to say about all this afterlife shit I had once considered nonsense? Yes, I could tell you about how some of my new friends seemed to defy the laws of the universe. Yes, I could tell everyone that I witnessed the impossible. But did that mean there

Introduction - WTF JUST HAPPENED?!

actually *was* an afterlife? And if it did, would that give me the thing I wanted more than anything in the world?

Most of us want to believe that consciousness can continue after bodily death. When my father died, I began to explore if that is even possible. But wanting to believe something, even with all of your heart, doesn't make it true. As a sciencey girl, I need evidence, and that's exactly what my story is all about.

Consider the words of Dr. Ian Stevenson, an esteemed psychiatrist who conducted years of research into past lives. "I'm not saying it's true; I'm just saying it happened."

I was raised in an intellectual "Christopher Hitchens-admiring," atheist family. We are culturally Jewish, but more of the "Jon Stewart-ey, love our bagels and lox" variety over the "pray in temple" type. I was taught that belief in God and an afterlife was at best wishful thinking, which brought out kindness in people, offering comfort during the darkest of times, and at its worst, a cultish form of religion, which spurred homophobia, sexism, and an anti-science mentality.

I was in my late 20s and things in my life were pretty good. I was launching a new startup, I had ended things with a toxic boyfriend, cleared out some harmful friendships, and I was meeting inspiring new people. Then, my father had a stroke.

'Of course, he will be fine.'

That's what I told myself. I was worried, but not too worried. But he was not fine. The stroke left him partially immobile before infections took over. My father had had me later in life, so he was not able to bounce back the way most of my friends' dads would have. And my world crumbled around me. My dad was one of my lifelines and a main source of my safety and support. Despite lots of fighting, (we are both strong-willed people), he was also a close friend.

While my mom, who was quite a bit younger than him, was in the midst of growing her psychiatric practice, he had already sold a

Introduction - WTF JUST HAPPENED?!

business and was able to dedicate time to his hobbies—reading, writing, playing poker, and me.

So, what does someone raised with no belief in an afterlife actually do when one of the people you love and rely on more than anything dies?

It was in the moment we got the news that reaffirmed just how much I turn to science for my answers about the universe. When the doctor told us that my dad wasn't going to be okay, but that "sometimes miracles happen," I felt as if I had flatlined right then and there.

"I'm a science person," I said, "not a miracle person." Until that moment, I hadn't even known how much that was true.

My dad was going to die.

I made a decision I would lie in bed and never get out—ever again. Somehow, in the middle of my deep grief, I discovered an unexpected new world, one that actually suggests evidence of an afterlife. This saved my life.

Please don't dismiss the possibility of an afterlife as wishful thinking. I grew up with a psychiatrist mother and her friends, including some neuroscientists, so I am as skeptical as you can get. If you think the way I did, it is terrifying to open your mind because you are already pretty positive that if you get your hopes up, even in the least, the end results can only disappoint and break your heart.

Giving the possibility of an afterlife a chance was by far the scariest thing I have ever done and I don't hold back from scary adventures. I traveled alone to Thailand, and with a friend through Bolivia. I moved to Paris where I didn't know the language to work in fashion for six months, but none of that compared to how scary it was to explore the possibility of an afterlife and sometimes it still freaks me out. But the alternative was worse. I couldn't accept the fact that my dad had been wiped from my life for good. I had nothing left to lose.

My exploration opened up a magical and beautiful world I had no idea existed. During my journey, I met some of the most incredible people I would previously have dismissed as self-delusional, or

Introduction - WTF JUST HAPPENED?!

worse, con artists. I experienced an awe I had never known before, spine-tingling moments and wondrous, unexplainable events.

I am not sure if I believe in life after death, but from my experience there seems to be an excellent argument for it, and trust me, I am more shocked than anyone to say that. Today, my life is rich with unexpected, mind-blowing occurrences. So, come with me now and follow this story, one that I could never have begun to fathom, where all I can say is, WTF? What the fuck just happened?!

1

How Hard Can Time Travel Really Be?

When my dad died, I lost one of the few people who cared for me unconditionally. I felt abandoned, betrayed, and isolated in my grief. Being an only child added to this pain. You are a parent's number one in the way you will never be anyone else's again. A husband can divorce you. Your friends can outgrow you. And your kids, no matter how much they love you, depend on you to care for them. They do not take care of you. Now I just had my mom. That parental, unconditional love was suddenly reduced by 50%.

I was shattered. I missed my father more than I even knew was possible.

Most of my friends had not yet experienced such a significant loss and no one knew what to say or do. I had no idea how I would begin to navigate this new world where I felt significantly less safe, less loved, less supported.

Memories of him kept circling in my mind. Memories? It was impossible to think about him as a memory. My father loved to read. He always bought me books, which we would have long in-depth conversations about for hours. We both loved to try new restaurants. The Four Seasons was his favorite. And he loved dessert. He would always ask for a small bite of mine, which of course was a huge bite.

"You just think I took a big bite because you are so small," he always told me when I complained. "I actually took a little bite, but it looks big to you." His defense stopped working by the time I was ten years old.

When he had been moved into hospice, no part of me believed this was really happening. How could I lose this vibrant, supportive, often-to-the-point-of-indulgence person in my life?

But the doctors had said they couldn't do a thing to change it.

My dad had always told me I could do anything. If there was a problem, he taught me to go solve it. So, I needed to DO something. I didn't believe in God and had no one to pray to. I did believe in medicine, but still nothing the doctors did had been able to save him.

What was left?

Science. I believed in science. When I started to think about it, I realized how science and technology can create actual miracles—inventions like airplanes, personal computers, cell phones and Zoom have transformed our definition of time and space. Not too long ago, those technological break-throughs would have been considered Sci-Fi fantasies.

One thing the Sci-Fi movies always showed was time travel. If science had turned these other Sci-Fi dreams into reality, could science show me how to time travel? I imagined ways my great-great-great-grandkids could come retrieve my dad, and eventually me, when civilization finally had DeLorean time machines. As long as my (future) kids had the date my dad (and all of our family going forward), got sick or died, and our locations, our future family could know when and where to grab us and drag us into a future when advances in medicine would save us.

I decided to Google Albert Einstein and Stephen Hawking, the smartest people I could think of, combined with time travel. I downloaded books and videos on Einstein's theory of relativity, read articles published by the top astrophysicists, and brushed up on my knowledge of black holes.

I had studied Einstein's theory of relativity in high school, but I had never realized its significance. That our conception of time, something we experience as linear, was not. That the laws of the universe could have a lot more to them than we perceive. To help explain his theory, Einstein offered a thought experiment about a pair of twins, one who traveled through outer space at close to the speed of light and another who remained earth-bound. When that traveling twin returned to earth, according to Einstein, they would be physically younger than their twin, as well as everyone else on earth.[1]

If the very nature of time was not as definite as I had thought, then what else was possible?!

The other beyond mind-blowing thing I discovered was what is referred to as Spooky Action/Bell's Theorem[2], which Einstein (as well as others) are also responsible for demonstrating. Because, on a quantum level, tiny atomic particles act REALLY weird, scientists remain baffled by this "spookiness." This was a huge insight into the fact that our consciousness seems to play a bigger and more inexplicable role on our physical world than I had ever imagined. Could our own consciousness actually affect how matter behaved?

Before going further, here is a bit of a high school science class refresher. Electrons are subatomic particles with a negative charge, protons are subatomic particles with a positive charge, neutrons are a subatomic particle with no charge and photons are tiny particles that transmit light. Protons, electrons, and neutrons make up atoms, which make up matter.

"Spooky Action" was discovered when scientists wanted to study whether electrons or photons behave as either waves or particles. To find out, they shot photons through a vertical slit on a screen to land on a wall behind it. When these photons passed through the screen, they behaved like a particle. That means that the photons passed through the slit as a single particle—creating a vertical line of dots on the wall behind the screen. When scientists shot the photons through a screen with two parallel slits, however, the photons then behaved like a wave, undulating like ripples in a pond. This showed up in the pattern created on the wall behind the screen. Since waves in ponds interfere and often cancel one another out, there were not the expected neat lines that particles passing through one of two slits would create. There was something called "interference patterns" when the ripples of the waves would "interfere" with one another.

Then it gets really crazy. When we observe which slit the photon goes through, tested by a detector set up to measure whether the photon travels through one or the other slit, the photon changes its behavior and acts like a particle. It *chooses* which slit it will travel through. When the detector is turned off, the wave pattern shows up. THE PHOTON ACTS DIFFERENTLY BASED ON OUR OBSERVATION! The very act of observing, or our own consciousness, affects the behavior of the most basic level of matter. And sometimes these photons "entangle," where the spin or behavior of one affects the spin or behavior of another. Change the spin of one photon and the other changes. To make this happen, photons have to communicate faster than the speed of light.

So apparently sometimes there are exceptions to the scientific laws of this world?

Scientists can't explain any of this. Nobody can.

Suddenly, my mind was open to the possibility that the things I previously knew to be 100 percent true were not so 100 percent known at all.

I WAS STILL GETTING PUMMELED. I never knew grief could be so brutal. So physical. As soon as my dad went into hospice, it was as if a fluid metal weight was settling into my stomach weighing me down, and pushing into my organs. It poured into my limbs, congealing and making it close to impossible to move them. When I did move, everything just hurt. It all escalated the night my dad passed away. When we got the call at 2 a.m., while I was huddling against my mom in my parents' bed, I felt as if my body exploded into heavy little bits all over the floor. I understood what it meant to feel shattered. This was not MY life. I could not identify ME in a life without him. Everything that had once seemed exciting—growing my career, falling in love, starting a family—now had an emptiness and sadness attached to it.

The only thing that made existence worthwhile were these wondrous hints that there was more than this material world. These at least gave me a little hope and curiosity.

So, I dove back in. I had had good luck with Einstein and time travel, so I Googled other names I trusted, combined with words like alternate dimensions. And crazy shit presented by trustworthy people came up. First of all, Stephen Hawking was talking in a serious way about shadow people reflecting into our universe from other dimensions?! When I thought of people talking about shadow people from other dimensions, I pictured either ghost stories at summer camp or late nights with my friends in high school and college, when we had drunk or smoked way too much of something. I was sure Stephen Hawking didn't mean it in the way people meant it when they told ghost stories, but his theory at least hinted that our world was not the simple mass and matter that we think we understand.

What else could be possible?!

Determined to find out, I streamed videos about the fourth dimension (fifth, if you are calling time the fourth) and the Tesseract, a mind-bending 4D cube that follows the fourth-dimensional laws of the universe. I watched a movie called *Flatland*[3] about a creature existing in a two-dimensional world, who then started to learn about our own three-dimensional one. This opened my mind to the possibility that we perceive only a tightly edited portion of what's out there. That was always partially obvious to me, in the sense that we don't know what exists billions of miles out in space, or that dogs hear noises we can't, but I had never thought how we might be seeing a minuscule portion of the reality immediately around us. This all offered a spark of hope, which in my darkness, I clung to with desperation. I knew I needed to be careful, though, not to draw any quick conclusions based on wishful thinking. I needed to remain neutral and tell myself I would accept whatever seemed to be true (or not true), whether I wanted it to be or not.

But even if it seemed nuts or completely unrealistic, I began to hope that any form of an afterlife might exist.

However much I tried to imagine though, I could not see how a person's consciousness could be anything more than a function of their living brain. Gone when a person was gone. But just because someone's consciousness was most likely the result of a particular set of firing neurons and brain cells, why couldn't that result happen again? While I might not be able to experience it as "Liz" and my dad would not get to experience it as "my dad," we could at least have an experience of consciousness as some future human being. We would have no ties to one another, or our previous selves, but we would still get to experience living. That option was not as good as actually being the "me" I am now, being my dad's daughter, but it was better than total obliteration. And maybe if this was the case, somehow, somewhere, someone had stored and now had access to a memory of a previous "me" or "them" that they had been?

I knew it was a long shot, but I went ahead and Googled, careful to word everything in the most scientific way possible: "Scientific evidence that people can live more than one time."

Then... this popped up: "Searching for the Science Behind Reincarnation on NPR" with the following description: "Jim Tucker, a psychologist at the University of Virginia, studies hundreds of cases like this and joins NPR's Rachel Martin to share his research on the science behind reincarnation."[4]

WHAT!!?! This was NPR, not some woo website.

Before clicking, I calmed myself down. I was sure this would turn out to be some new-agey crap. Or a Sci-Fi theory, true in concept, but not realistic, just like time travel. I took a breath and clicked and read and then... I discovered something, that if true, hinted that our entire scientific understanding of life, death, and consciousness was wrong.

A reasonable, scientific-minded psychiatrist and professor at the University of Virginia, Dr. Jim Tucker, was studying cases of kids with past-life memories. AND he seemed to be getting valid results? From his interview, I learned that Dr. Tucker had a mentor, Dr. Ian Stevenson, who also had been a psychiatrist and worked for the University of Virginia School of Medicine for 50 years before he passed away in 2007. He had even been chair of the school's Department of Psychiatry.

I kept trying to calm myself down since this must have a catch, but my heart pounded. Past life memories were being seriously studied by psychiatrists and professors at a major university?!

I immediately downloaded Dr. Tucker's first book, *Life Before Life*.[5] My heart pounded as I began to read. Waves of tingling ran up my body, poured through my stomach, and paralyzed me for a second. It felt like when an unexpected twist ending in a movie is revealed, but much more intense since this was real life and I had a lot more at stake. I got Dr. Tucker's next book and a few books by Dr. Stevenson and devoured them all. These two psychiatrists used a detailed scientific method to carefully study children from a variety of cultures and belief systems who claimed to have memories of previous lives. They had accumulated thousands of cases and they never talked about karma or good deeds or tried to make anything fair or work the way we would want. It was facts—just facts. Well researched and carefully studied facts.

One thing that stood out about these kids' stories was the everydayness of their memories. Nobody was a hero or royalty. They remembered things like leaving their backpack on the table when they got home from school, or a small neighborhood candy store—things we really do remember when we think of our childhood. I found websites where people shared their own past-life memories. Ninety percent of them (aside from the few expected nuts) matched the tone and patterns of Dr. Stevenson and Dr. Tucker's research: Kids would stop speaking of these memories around age five. There were intense emotions attached to the trauma of their past-life deaths, and feelings of love and longing were attached to the "memory" of their previous families. Like the kids in Stevenson and Tucker's study, they recalled everyday things from their past lives in a steady and factual tone.

What the fuck was this?!

I had to share this discovery with my mom, a skeptical and logical, "consciousness-is-a-function-of-neurons-and-belief-in-an-afterlife-is-wishful-thinking" psychiatrist.

Me: Mom! You have to come check out this person Doctor Jim

Tucker. And his mentor Ian Stevenson. They are actual psychiatrists. And they seem to think there IS evidence of past lives.

My mom had been handling my dad's death differently than me. While I was ready to spend the rest of my life in bed, trying to discover the answers to the universe's greatest puzzles, my mom was not so interested in trying to turn back time or learning about other dimensions. After a week, she was out of bed and back at work.

> **Mom:** There have always been people talking about helping people uncover some past life memory. It was big in the seventies in Greenwich Village among really lost people.
> **Me:** MOM!! I know that kind of thing. This is different. Really. Doctor Tucker is a child psychiatrist.
> **Mom:** Unfortunately, the mental health field can attract some flakey people. Not all are licensed, and they can just say they are therapists.
> **Me:** He is an actual psychiatrist, and works at UVA teaching psychiatry. NPR did an episode on him and Scientific American wrote an article. Even Carl Sagan said that his mentor Doctor Ian Stevenson's research on kids with past life memories was worthy of further studies. And this is the craziest, JAMA, The Journal of the American Medical Association, even said reincarnation is the most likely explanation for some of these kids' past life memories!

I showed her the articles and websites and books.

> **Mom:** UVA? Child psychiatrists? Hmmm. I would be curious to hear what he does say. But there are a lot of people studying out-there theories.

She did not sound surprised and still very much expected to find the catch.

> **Me:** When you start looking into their research, though, it is pretty

astonishing. There really seems to be something inexplicable going on.

My mom may have been skeptical to protect herself (and me) from any wishful thinking and ultimately crushing disappointment during such a vulnerable time. But I wasn't so cautious. I dove in, full force, full of curiosity to see if there really was any valid hint of an alternative to my deeply painful new reality.

2

I'm Not Done Talking About Reincarnation

This research of past lives gave me enough hope that I had the motivation to leave the house and meet a friend. A close friend I felt safe with.

When I stepped outside, the world felt surreal. So, this is the world without my dad in it. This is what it feels like to be me in a world without him. I was now a different person. I was someone who had lost a parent. I could go to his hangouts, even scour the entire globe, and he would not be there. And he wouldn't be there tomorrow or next week or even next year.

The world felt like a strange dystopian universe, bright and sunny with cheery people walking around in the midst of tragedy and despair they did not see or pretended not to see. And every inch of my body hurt. A lot.

I walked past my old nursery school. I remembered how my dad would push me down the block in my stroller. "Faster! Faster!" I would giggle as he raced down the block faster and faster, but probably nowhere near as fast as it had felt at two years old. I now hurried past the school before the kids came out and I would have to see them being picked up by their moms and dads.

Despite wanting to cocoon back under the covers, I continued on to meet my friend Christine. She was one of the few people I knew and respected who was into "new-agey" stuff. She gave me a huge hug and told me that sometimes life is just shit. I decided to tell her what I had been exploring.

> **Me:** Warning. This will sound so weird. I found these scientists who say that we live more than one life.
> **Christine:** Sure. I always thought that.

I told her the rest. She listened, then she shared her experiences of meditations, where she had visited other dimensions. I can't say I considered it valid, but I listened in a way I never had. Maybe? Maybe that hinted at something?

Then I opened up further.

> **Me:** But I... I have to be honest about something. I don't know if I want to do this anymore. Be here anymore. I just hate the world without my dad. The rest of my whole life will be second best, even on the best days. I can't purely enjoy anything again. And I haven't just lost him. I lost my mom, too, because she is so depressed and transformed. And I lost me—because I don't know me and my life without him. I honestly don't like my life anymore. I can't see liking it again. I... don't freak out... but I just want to die.

I will be forever grateful for Christine's non-hysterical, calm response. Some part of her must have known I didn't really mean it (even if I didn't know that at the time) but I had no other frame of reference to express how awful I felt.

> **Christine:** You know what? That is your choice. And I respect it. But I think you are an amazing person and you have so much going for you. There are people who I would say, "I think they should go ahead," but you are exceptionally kind, and it is rare to find truly good people. So, I think that would be a waste. Why not

give it a year, then if you still want to, go ahead. It will always hurt, and you will always miss him, but you don't know yet if it will be worth it to stay alive or not. You never tried to live in a world without him.

True.
I listened quietly as Christine continued.

Christine: I know you feel like everything is over, but I actually think you are at the start of this incredible journey with all you are exploring. Like when a movie opens at the very beginning with this huge dramatic and often traumatic event that the character thinks is the end of everything, but it turns out to be the beginning.

AFTER MEETING WITH CHRISTINE, I went back to my parents' place—I could still call it that, right?—And climbed back into bed. What Christine said stuck with me. I was too curious about the results of my research to back out now.

I thought back to a night towards the end when my dad was in hospice, when I was too scared to sleep. Or actually, the sleep part would have been fine, but I was too scared of waking up and remembering all over again. I had been lying in my parents' bed with my mom under the same brown sheets and cream blanket that they had had since my early childhood. The same small Oriental rug and Chinese antique vases on brown wooden dressers were still there. I remember thinking how strange it was that our home, witness to our entire life together as a family, was completely oblivious to what was happening. Shouldn't it have reacted… somehow?

Mom: Sorry I could never raise you with any belief in God. I'm so jealous now of people who believe. But if I had, I would have felt as if I was lying to you.
Me: I would have seen through you.

Mom: I know. You have always had a way of seeing right to the truth of things. It could be a major pain in the ass when you were little, but I'm glad you are that way.

I wished I wasn't. I wished I could be the kind of person who believed. No evidence needed.

Now, here I was discovering a possibility, however minuscule, that we could get just that. Not a god or anything, but that is not what we cared about. I only cared about seeing my dad again, no matter the terms.

I was scrolling through Netflix to find something that would take me anywhere but where I was, when I saw the following blurb about a movie called *Wake Up*[1]: "Jonas Elrod was leading an ordinary life until he woke up one day to a totally new reality. The documentary follows this fascinating story of an average guy who inexplicably developed the ability to access other dimensions."

I called in my mom in to watch it with me. With a sigh and a slight eye roll, she agreed. Burrowing under the cocoon of covers on her side of the bed, she took two of the four pillows and settled in next to me. Her side on the left, my dad's side on the right. Me in between, when I had been much smaller and they let me. I rested my head on her shoulder. That had often eased the ache when I was sad. This time it did nothing.

In the documentary, Jonas Elrod is a typical, slightly hipster, 20-something guy, living a pretty normal life until the sudden loss of a close friend prompts voices in his head telling him to, "Spread the word. We don't end after we die." He also started seeing what he could only describe as ghosts. A psychiatrist couldn't find anything wrong with him. My mom even knew who the psychiatrist was and he was considered serious and was well-regarded! I looked over at my mom's face. She looked genuinely baffled.

From there, Jonas went on a journey to make sense of all of this. He met some remarkable people along the way. A few of the experiences and people he encountered were exactly what I expected, such as an ashram spiritual center that had bathrooms with no doors where everyone could just… watch?

And a few were nothing like I expected. There was Dr. Gary Schwartz, a Harvard-educated former professor of Psychiatry at Yale and current professor at the University of Arizona. I later Googled him and learned that he conducted scientific studies on mediums. So, a scientist was actually taking people who claim to talk to DEAD PEOPLE seriously?

Another fascinating person Jonas Elrod introduced us to was Dr. Roger Nelson, a professor of science and part of Princeton's (Princeton! As in the Ivy League, very credible non-woo Princeton!) Princeton Engineering Anomalies Research (PEAR) Lab, which studied "non-normal" views on consciousness and how the mind can affect matter from 1979 to 2007.[2] Dr. Nelson demonstrated something called a Random Number Generator (RNG), a machine created to randomly generate 1s or 0s. It generates these numbers from a process that does not use an algorithm, making it truly random and completely unpredictable. When people who were brought in as test subjects, mentally focused on having the RNG show a higher number of 0's over 1's for example, more 0's would show up in statistically significant numbers?! Weird!

Dr. Nelson is currently working on something called the Global Consciousness Project,[3] an experiment where a group of Random Number Generators (they originally started with around 11 and are now up to about 70) are placed around different parts of the world. They run 24 hours a day and send data to a centralized location. As stated, these generators usually show the randomly expected pattern of 0s and 1s, but when there is a global event that gets a large percent of the population's attention, the patterns shift, and the results seem to show no longer random patterns of activity. Instead, the number patterns move from random to orderly with higher and organized groups of 1s and 0s. It moved past the expected random number pattern to a more orderly pattern during Princess Diana's death, and on New Year's Eve 2000. The craziest of all was on September 11, 2001, when it not only moved from random patterned results to orderly results, but started to react and change BEFORE the tragedies occurred. Really, really weird!!

As Jonas Elrod met these people, he remained grounded. He

was not a "believer" in anything, and some of the things he encountered, if true, were absolutely without rational explanation. Feeling a tiny twinge of hope, and noticing a flash of—was it also hope?—on my mom's face, we snuggled under the covers to escape. Her into sleep and me into the possibilities.

3

A Class With Someone Who Believes All This Afterlife Stuff

I was sitting with my mom in our kitchen at a small round wooden table below the Cézanne "Dish of Peaches" painting. She was in the same chair she's had for as long as I could remember, and I was in the chair I always had sat in since I graduated from a highchair. My dad's chair was empty. My mom looked over at me with an odd combination of bafflement and a kind of sympathy reserved for the very gullible.

> **Mom:** Well, you won't believe it but apparently Doctor Berger knows Jim Tucker.

Dr. Berger is a psychiatrist in my mom's network whom she respects.

> **Me:** KNOWS Jim Tucker? What do you mean knows? Like read his books or knows knows?!
> **Mom:** Knows personally.
> **Me:** WHAT? Is he normal? Is he sane? Did you ask her? What did she say about him?

My mother set down her cup of coffee.

Mom: That he is smart and sane and very nice.
Me: Oh my god!! Ask her. Can I meet him?
Mom: I am not sure I can just ask for her to have him come hang out.
Me: Get him to do a talk at your institute!
Mom: No.

I knew there was no negotiating that one. Under no condition would her institute have someone come talk who studied anything to do with past lives.

Mom: But... Doctor Berger asked if she could hold a series on some parapsychological research and because of you I said okay.
Me: REALLY! Will you go with me? Please?!

My mom begrudgingly agreed to go, but she made it clear she was doing it as a big favor to me.

So that was one of the hardest things I was noticing about grief. That my mom and I, who were both grieving the loss of my dad, had very conflicting needs. I wanted her to be excited about Dr. Berger giving a talk. I desperately needed to talk over everything I was studying with her. My mom is rational, logical, and healthily skeptical, and I craved her opinion to see if I was genuinely onto something or if I was falling victim to delusion or wishful thinking. But she did not want to go there.

When Dr. Berger's talk started, I listened intently and took detailed notes, even though I felt like I could barely move. Grief was so *heavy*. It was hard to sit upright in a cold folding chair in this fluorescent lit room. My body hurt. My head hurt. Everything hurt.

Dr. Berger mentioned William James, the brother of author Henry James, and the experiments he had conducted back in the 1800s on a psychic medium named Leonora Piper. William James himself had not believed any of this could be true, but was so intrigued when he saw her do a reading for some friends of his, that he was driven to investigate and find the catch. He had taken proper skeptical precautions such as carefully hiding the identity of who Leonora Piper was going to have a session with until the reading started and having spies follow her to see if she was trying to cheat and get any information in advance. William James never found a "catch."

Was there actually something to mediumship? Did mediums still exist today and take clients? How would I even find one?

Dr. Berger brought up something called remote viewing, or the ability to see areas and locations you were not at physically. She mentioned Ingo Swann as someone who was apparently very accurate at it. Remote viewers like Swann would close their eyes and travel to or psychically view these places. Controlled experiments were conducted and some people accurately described details of the location significantly beyond chance!

In another talk, a well-respected therapist, Dr. Kleinman, spoke about his experiences working with patients in India, a culture where it was widely accepted as fact that our spirit? Consciousness? Lives on after bodily death. He told us about a patient whose grandmother passed away. The grandmother had collected statues of a rare bird in Jaipur, where they lived. When the woman went home from her grandmother's funeral, that species of bird was sitting in the doorway of her grandmother's house. In Indian culture, even in the metropolitan cities such as Jaipur, this was assumed (without question) to be this woman's grandmother. According to Dr. Kleinman, his patient was a successful, modern businesswoman, living in the heart of urban Jaipur. She was well educated, well-traveled, and not one prone to "village superstitions."

I began to think of the arrogance of blowing off, without even a consideration, an entire culture's experiences and worldview.

Dr. Berger shared a few more remarkable experiences: there was a patient at a hospital where she worked with multiple personality disorder. One personality had a deadly allergy (actual, not in his head) to nuts and the other one did not. If the allergic personality ate nuts, he would end up in the hospital and possibly dead, but if the other personality *in the same body* ate nuts, there would be no physical reaction. It brought me back to the Einstein experiments about matter changing under certain circumstances. Could this be tied to the "observer" phenomenon? Yes this all went against my entire understanding of how the world worked, but it was no more shocking than the fact I would never see my dad again. But this time it was a good kind of shock.

Dr. Berger brought in another speaker to talk about research on the influence of thoughts on physical matter that had been conducted at Princeton University's PEAR laboratory—Princeton! I still could not believe this had been done at Princeton!—The experiment used the same Random Number Generator I had learned about in Jonas Elrod's movie. The machine that randomly produces a 1 or 0. It is expected to generate a

50/50 (or reasonably close to that) percent of 1s and 0s. And if you remember from the movie, when the people who were test subjects tried to use their minds to influence the results, the number they were thinking of appeared significantly more than 50 percent of the time.

The researchers noticed that people who meditated did better and, once people started doing really well, i.e., achieving results way beyond the odds of chance, the results would stop. They could only defeat the 50/50 odds for so long. One theory as to why this occurred is that a part of the mind does not believe this is possible (understandably), so when someone set a target (either the 1 or 0) they would beat the odds very quickly, and then tell themselves it wasn't possible, and the results would stop. Interestingly, when the subject gave up and stopped caring, their results soared again. We seem to have a need, in our real lives, too, to stop ourselves before we go past what we have been told is possible.

When the results of this study were shared with mainstream science, the response was less than enthusiastic. Dr. Berger told us that a respected scientist replied, "Even if it were true, I wouldn't believe it." That response went against everything I had ever thought about scientists and their neutral assessment of facts. How much were we missing by not noticing things if they didn't fit into our system of belief?

What if I were to let go of any and all beliefs about how anything worked and not tell myself certain things were untrue or impossible. What if I just followed and noted what I observed no matter what, with no preconceived notion or opinions?

4

A Few More Books Get Me Thinking This Isn't All Batshit

I missed my dad. I missed me. That he was gone felt more real if I participated in the world, so I decided to not participate. All I did was search for books to read and videos to watch about this astounding possibility and intriguing hints that the universe was different than I thought it was.

Twinges of excitement would break through the molasses of grief at the vast possibility of these phenomena and discoveries of this new world. My dad had never been in this alternate world I was learning about, I had never been in that alternate world, we had never been in it together. I therefore did not feel his absence when I dove into it. At least not as much.

In my escape from the material world, I found four books that opened up another level of this journey, all hinting that how I "knew" the world worked, was not necessarily the whole picture.

The first book was called *Spooky Action at A Distance* by George Musser.[1] As explained by Musser, "spooky action" is the entangled particles and the double slit experiments of Einstein that I had learned about at the start of my research. I typed this quote from the book into my phone: "And if it does turn out that space and time are the products of some deeper level of reality, who knows what new phenomena await our discovery?"

The next book I found was *The ESP Enigma: The Scientific Case for Psychic Phenomena* by Dr. Diane Hennacy Powell.[2] This woman came from a scientifically educated family. She studied at Johns Hopkins University School of Medicine, then taught neuropsychiatry at Harvard Medical School. Here was this very logical and brilliant woman making a case for psychic abilities. When I read that "one of the reasons psychic phenomena have not been more widely accepted within neuroscience may be that most neuroscientists have not investigated the existing data,"[3] it gave me further hope. This quote backed up what I had been noticing regarding how the skeptics responded to Dr. Jim Tucker. Every criticism I read about his work was very surface-level and was always addressed if you actually read his books, which many of his critics had proudly not even done.

I made sure my Mom heard that quote by Dr. Hennacy Powell, although I don't think she gave it very much credit based on her generic "Oh, okay." reply.

I then jumped into exploring String Theory. I knew it was based on the theory that our universe is made up of multi-dimensions of vibrating strings, but had never examined it in depth. I devoured *Warped Passages* by another well regarded scientist Dr. Lisa Randall,[4] a theoretical physicist working in particle physics and cosmology, and a Professor of Science on the physics faculty of Harvard University. *The Complete Idiot's Guide to String Theory*, another book by George Musser,[5] was written more simply and helped clarify the scientifically written *Warped Passages*.

If traditional Stephen-Hawking-approved-science knew that our universe was made up of many dimensions—dimensions that we 3D humans could never even perceive—what other states of consciousness could exist? Could there be states of consciousness that did not need a brain or body?

Another book I chose to explore was *Rethinking Immortality* by Dr. Robert P. Lanza.[6] Yet another brilliant scientist who considered the survival of consciousness a possibility and made a great argument for it. Dr. Lanza is no fringe scientist. He was a Fulbright Scholar, one of the leading stem cell scientists highly regarded in the scientific community. He was also the first to clone an endangered species, was featured on Stephen Hawking's Stem Cell Special, and was part of the team that cloned the world's first early- stage human embryos. In his book, he quotes from Ray Bradbury's book *Dandelion Wine:*,[7]

> *"Imagine all the days and hours that have passed since the beginning of time. Stack them on top of one another like chairs, and then seat yourself on the very top. This is how the human brain tends to caricature time. The statistical probability of being on the top of infinity, the mathematician might add here, is so small as to be meaningless."*[8]

The profundity of this quote from his book blew me away. Here was another challenge to the meaning of time.

Furthermore, we only know what five percent of the universe is; the rest is called dark matter and we have no idea what that actually even means. Add to that the relativity of time, the fact we have no idea what would happen if we traveled into space in one direction for one million years, what everything will look like in 100 trillion years, or who was the first conscious human. The endless possibilities that all of this implied helped me realize how vast and filled with secrets, twists, and surprises our universe is. We know nothing. We cannot say we know.

This exploration of the "great unknown" and the surprises I could discover along the way gave me a hope to go on.

5

The Sacred Scriptures Of Gary And Julie

Who was this Dr. Gary Schwartz, the one from Jonas Elrod's movie who conducted studies on mediums? I downloaded and devoured his books as well.[1] Dr. Schwartz holds a Ph.D. from Harvard and had been a professor of psychiatry and psychology at Yale University. He is currently a professor at the University of Arizona and the Director of its Laboratory for Advances in Consciousness and Health.

My heart was in my throat with hope—could this be real? As I dove into Dr. Schwartz's books, I kept getting more and more mind-blown. He was conducting a variety of scientific studies with mediums that were getting inexplicable results. These mediums—these human beings—seemed to really be communicating with deceased people. How in the fuck?

The thing that made my heart sink though was that there were a lot of questions (some more accusatory in tone than others) regarding holes in his procedures and protocols. Did that mean his whole studies and hypotheses were… invalid? But with innovative science and research other scientists often pick up where one left off and tighten the holes and protocols? Right??

What else was out there?! Was anyone else doing this? I then googled "scientific experiments with mediums" to see if so, and I came across someone else called Dr. Julie Beischel who had taken the foundations of Dr. Gary Schwartz's ground-breaking studies and experiments as a base. She and her husband, Mark Boccuzzi, then went on to found The Windbridge Institute, which conducted a variety of studies with mediums for years! I downloaded one of her books, *Among Mediums*.[2]

I was impressed with Dr. Beischel because she took the same approach I was trying to follow by studying just the facts and what she observed. One of the opening lines in her book was: "Now, what if we calmed down, put aside our assumptions about how the world works, and actually applied the scientific method to the phenomenon of mediumship? Well, I did just that, and this book reviews what I discovered."[3]

Their research became my main lifeline for the next phase of my exploration and emotional well-being. (Or if not actually "well" being, at least a lifeline in minimizing my suffering.)

I spent about two months doing nothing but reading Dr. Beischel's books and the studies on The Windbridge site, and Googling all about her. Anytime I wanted to say something (which wasn't very often), every single thing I had to say began with "According to Doctor Beischel" or "According to this study I found on the Windbridge site," followed by some sentence that was beginning to give me hope of an afterlife. Although I did mix it up some with quotes from Dr. Jim Tucker and Dr. Ian Stevenson as well.

I was probably kind of annoying. My mom semi-indulged me and semi-rolled her eyes.

Dr. Julie Beischel approached everything with a scientific and logical mind. She stayed away from words such as "God" and "Heaven," focusing only on data from her experiments. She had received her doctorate in Pharmacology and Toxicology from the University of Arizona in 2003 and had walked away from a lucrative career in Pharmacology to scientifically study mediums full time.

This change in her life path was inspired by the loss of her mom and an experience she had with a psychic medium she went to on a whim. This medium knew stuff about Dr. Beischel that she could not have known normally. Dr. Beischel was intrigued. Coupled with her deep curiosity, this experience inspired her to start The Windbridge Institute, where she conducted a number of highly intriguing triple, quadruple, and even quintuple blinded studies of mediums.

Logical, highly educated, and open-minded scientists studying mediums and the potential of an afterlife, with a neutral mindset and getting results in favor of both? What the fuck?!

In The Windbridge Institute's various levels of experiments, mediums were selected to give a reading to communicate information from a person who had passed away, (referred to in the field as a "discarnate,") to one of the discarnate's loved ones who was still alive (referred to as "a sitter"). To control the experiments, the mediums were given no advance information about either the discarnate or the sitter. At times the mediums were on the phone with the sitters and at times they weren't. Sometimes the medium would hear the sitter's actual voice, who would only answer "yes" or "no" or "I don't know." Sometimes the sitter would push a button for yes, no, or I don't know, so the medium couldn't even hear the voice of the sitter and possibly pick up cues.

What absolutely blew me away, was that these studies were taken up to a quintuple level of blinding. I had to read this study multiple times (as well as many others) to make sure I actually understood and was not missing something, because what the studies were stating was completely defying the laws of the universe as I knew them to be—while using highly scientific protocols!

Here's a description of Windbridge's protocols for their quintuple-blinded studies: While there were many mediums that participated, let's take one and call her Susy. So Susy knew she was going to give two separate readings, during which she would bring in one discarnate to each of the two sitters. Susy was given no information, aside from the first name of the discarnate at the time she was to give the reading. In the first experiment, Susy never meets, sees, or hears the sitters. Susy gives the information she is getting for these unknown sitters from their deceased loved ones, to a proxy—an experimenter who was also blinded to the identity of the sitters in that reading.

The two sitters in this experiment were intentionally matched to be read by the same medium. The reason for this "pairing" of the two sitters (who never meet) is so that when they score Susy's reading for accuracy, they also score a control reading, that was done for the other sitter. To be considered an appropriate "control" reading, each sitter needs to be "paired" with someone who has lost a person of the same gender and with a name that would be from the same cultural background. Their discarnates also need to be different enough that the two could distinguish the readings.

This meant that Susy would be reading both sitters' discarnates at two different times as two completely separate readings. Both sitters would receive the write up and notes from both readings and not know which was meant for themselves. They were to score both readings and then state which one was their own discarnate. This prevented sitter bias. For example, if the sitter was desperate to hear from their loved one and for mediumship to be true, the sitter might score too positively. This extra reading added a control.

The experimenters were blinded also. For example, the first experimenter (let's call him William) would meet with the sitters separately, train them, and determine the sitter pairs. William did not know which medium would read which pair of sitters, or which of the two readings was meant for which sitter.

Then there was a second experimenter. Let's call her Lisa. Lisa would sit on the phone with the mediums, recording the readings as they came in for each of the sitters. The medium again, has only been given the first name of the discarnate that they are "bringing in" for the session. After taking the information, Lisa would write out the details from the reading and properly format it for the sitters. But… Lisa would know only the name of the discarnates as well. She would know nothing about the sitters, who they are, who the readings are for.

Then there is a 3rd experimenter. Let's call her Jen. Jen then interacts with the sitters while they are scoring their readings. She sends the blinded-paired-readings to the pair of sitters, but does not know which medium performed which pair of readings. Jen also knows nothing about the discarnate or which readings came in from which discarnates for which sitters.

I was stunned.

The results of the quintuple blinded studies (as well as other studies of various levels of blinding) were showing significant results: that yes, mediums were getting accurate information beyond the odds of chance. Some of the mediums ended up scoring WAY above chance with the information that came in. Remarkably, some communicated information about the discarnate to a sitter with 90 percent accuracy. The more impressive mediums relayed detailed, unknowable information to sitters about their discarnates, such as nicknames, personality traits, hobbies, careers, and verifiable events from their lives. The test did not award points for generic information, such as a discarnate saying how much they loved or missed a sitter.

Were they telling the truth? Did these experiments really happen this way? If so, this was the MOST paradigm-shifting thing I had ever encountered. Why were people like Stephen Hawking not all over this? There had to be a catch.

If Dr. Beischel's experiments were true, did this mean I could actually talk to my dad again?

MY DAD always had this vibrant and often inappropriate sense of humor and did not hold back saying what he really thought. He always said what everyone else was afraid to say, which ended up putting everyone (aside from the most uptight) at ease. I missed that.

One of my favorites was during our Passover Seder. While we never believed in, or really even thought about God, we loved the fun and tradition of the holidays. We celebrated both the Christian ones (what my mom grew up celebrating) and the Jewish ones (what my dad grew up celebrating).

Dad: Why would the Egyptians ever have Jews build the pyramids? I would much prefer to have had the stoic WASPS.

Raised a WASP, my mom would laugh and roll her eyes.

Dad: Can you imagine? Oy my knee! I can't carry all of those heavy bricks. I'll throw out my back.

And he backed up that Jewish stereotype in real life by calling our doorman to do everything, from changing a lightbulb to figuring out how to turn off the alarm clock whenever those perceived emergencies came up.

I wanted that humor in my life again.

I went to The Windbridge Institute website and scrolled through the alphabetical list to find mediums in New York. Although I had read that some mediums did readings over the phone, which was how Dr. Beischel conducted her studies, I wanted to sit with a medium in person. I wanted to watch what the medium was doing and what they weren't doing.

The first name on the list of New York-based mediums was Laura Lynne Jackson. Before I contacted her, I did everything I could to protect my identity. I created an email address without my name or any other identifying information in it. I installed a VPN onto my laptop to ensure she couldn't trace my IP address back to me. Because the sitters in Dr. Beischel's studies revealed only their first names, I decided it was probably best to follow that protocol and use my fairly common first name.

A few days later, I got a response telling me there was a two-year waitlist to sit with Laura. Ugh she was probably just a charlatan anyway! And how many dumb people were there in this world to require such a long waitlist? I put her email to the side to consider whether it was worth spending my time and money for a probable disappointment.

After mulling it over, I emailed Dr. Diane Powell, the author of *The ESP Enigma* and an authority on psychics. I sent it from my new fake email, mainly because I was embarrassed to be asking a scientist about mediums, even one who believed there was more to the world than "normal" material science.

> **To: Dr. Diane Hennacy Powell**
> **From: Liz**
> **Subject: Mediums/psychics**
> Dear Dr. Powell,
> I have become a big fan of your books and research. I am a bit new to this world and am looking to find a psychic/medium who is respected in the science world as not a fraud. I am in New York but if someone has proven to be very valid and is in another area, I am willing to travel (if needed).
> Thank you so much for your time.
> Liz

I didn't expect a reply, and even if I got one, I expected her to tell me there was no conclusive data yet for any mediums. Instead, I received the following email:

> **To: Liz**
> **From: Dr. Diane Hennacy Powell**
> **Subject: Re: Mediums/psychics**
> Dear Liz,
> Of the mediums and psychics I've met, the one who has impressed me the most is Laura Lynne Jackson, who is both a medium and psychic.
> Best,
> Dr. Powell

That same person?! And Dr. Powell didn't dismiss this all as nonsense.

I sat on it for about another two weeks, digesting this information before I logged back into my fake email, logged into my VPN, and I copied and pasted and re-sent the same awkward email I had sent to Laura before and I got on the waitlist. The two-year long waitlist. So, two years of waiting in suspense to see if there could be any truth to this.

A few weeks later, I went with my mom to a book reading at her Psychiatric Institute. The book was about Mark Goldman, a psychiatrist my mom knew before he passed away. His wife was there.

His wife gave a warm hello to my mom. Her face glowed as she excitedly shared:

Mrs. Goldman: Guess who visited me last week? Mark! He started playing with the electricity of my clock. I recently went to this medium who not only knew his personality but told me he would be playing with electricity to show me he was there.

My mom smiled politely, but I couldn't contain my enthusiasm.

Me: Whaaat!
Mrs. Goldman: Oh my god. Look at your daughter—she is really intrigued!
Mom: Actually, my husband, her father, just passed away.

It still felt like a stab in the gut to hear. Too real!

Mrs. Goldman: Oh, I'm sorry to hear that. Would you be interested in this woman's information?
Me: Umm—Yes. Thank you. Yes, I would.

Thankfully, my mom held back her eye rolls.

Though this medium wasn't certified by the Windbridge Institute, I went home and drafted an email to her from my fake email address.

She replied within a few hours. She told me a potential sitter apparently had to wait at least six months after a person's death before trying to contact them through mediums. She explained that the recently deceased need this time to adjust to their new existence on "The Other Side" and to summon the strength necessary to communicate with the living.

That sounded like a pile of bullshit, but contacting a medium in the first place sounded like a pile of bullshit.

I booked an appointment for three months down the line.

At this time, I was in the heart of Dr. Lisa Randall's book on string theory. It was one thing when a man wearing a tinfoil helmet discussed other dimensions, but quite another when a Harvard professor and expert on theoretical physics working on particle physics and cosmology research did.

I also continued to visit Laura Lynne Jackson's website.

"Don't become believey! Or fall into some wishful thinking trap!" I reminded myself as I noticed she too had a book. I hesitated then thought, "Fuck it. If I'm gonna do this, I'm gonna do this." I downloaded her book, *The Light Between Us*.[4]

To keep myself from feeling too "woo-woo," I alternated between Laura's new-age-y book and Dr. Randall's scientifically sound research.

Laura's book wasn't at all what I expected. It was grounded and well-written. She struck me as intelligent and relatable. A mother of three, she was a former English teacher and didn't try to sell anything, neither a product nor an agenda. I liked her, and her book. It covered some of her more remarkable readings, which were fascinating but obviously impossible to verify. What also impressed me—and gave me more hope—was Laura's work with a psychologist named Dr. Jeff Tarrant, who used an EEG to detect and monitor the electrical activity in her brain during readings. Her brain waves during readings, according to Dr. Tarrant's tests, were equal to someone who had suffered a traumatic brain injury?!

Even more interesting was the fact that the electrical activity in her brain lit up in completely different regions when she gave psychic readings (when she would read information about a living person) versus a medium reading (when she would communicate with a discarnate). I had no idea there was even a difference between a psychic reading and a medium reading.

I still didn't know if she was just tricking people or was deluding herself into thinking she was doing something genuine, but these results seemed to show there was something more than either of those explanations.

Laura described her test for The Windbridge Institute, giving a medium's perspective of the experience. She also had participated in a similar test for the Forever Family Foundation, another place that examined mediums, though her tests with the Foundation were conducted in person, while all of her tests with Windbridge were held on the phone. Still, to score highly on both sets of tests, she needed to communicate accurate information about people who had passed away that she had never previously met. While she could have made up the details about personal clients for her book, it would have been much harder, if even possible, for her to cheat on those tests.

Laura also wrote at length about seeing cords of light and energy, which she claimed connected us to one another and to our loved ones on The Other Side. Was this idea any weirder than string theory? Or quantum entanglement? Maybe her brain worked in such a way that she saw quantum entanglement connections as cords and that was the way we all connected to each other—through energy particles that got entangled.

I was lying in the dark on my parents' couch pondering this idea when I fell asleep and began to dream. I tend not to remember my dreams, but this one felt so real. I don't think I had ever experienced a dream that felt this real although the content was far from realistic.

I was in a kind of outer space, floating in endless black and surrounded by stars. Coming toward me in a tunnel of waves was this shooting golden light. A blond woman was standing in the middle of the light, which because of their shared hair color, I recognized as Laura and Dr. Randall, something only possible in dreams.

Pulsating out from a central point, the light waves vibrated up and down and I could see and feel them move around me, somewhat like the physical sensation when you are riding over waves on a stormy day in a boat. But, instead of feeling as if I were riding on them, I felt as if each wave pulsated and rhythmically rolled through me. It was much more powerful than any physical sensation I had ever had in a dream. It felt like a real physical sensation, although I had never actually experienced a sensation exactly like that before.

I understood that the waves were originating from a set of dimensions made of tiny strings, which vibrated in a musical rhythm to a universal HZ frequency, which rang throughout the multi-dimensions described in string- theory, originating at a core center of the universe.

Caused by the microscopic strings of string theory, these vibrating waves were the source of the same vibrating cords of light and energy that Laura said connected all of us to one another across dimensions—including to our loved ones on The Other Side. It was as if Lisa and Laura were both expressing the exact same universal truth using two very different languages, approaches, and frames of reference. I understood this not through words but through what I can only describe as some transfer of knowledge and through the experience of these vibrating waves.

This incredible feeling of warmth, happiness, and clarity all came to me as these golden waves rolled through me. A final pulse shot through me and I woke up with this feeling of a vibrant glow and calm lucidity. My brain and body were hot and tingly, and I felt as if I were still moving. A serene and edifying thought stayed with me as I lay recently awakened from my dream state: 'So that is how this universe works and how the spiritual explanation of mediums ties in with the most up to date scientific research.' Of course, I returned to my normal skepticism tucking this away as most likely just a dream. But still… ?

Not long afterwards, I was again cuddled under a pile of blankets on my parents' couch, one of the places I retreated to when I was a child to sleep or hide whenever I was upset about something that felt like a big deal at the time. Sometimes, I returned there as an adult to lie huddled in blankets when I needed refuge from adulting.

My dad would come in and awkwardly stand over me for a second trying to figure out what to say. Usually, he would just say, "Are you okay?" or "Do you need anything?" Knowing he was there was soothing, even if he never did know exactly what to say.

So there I was, burrowed under the covers, when I looked up and I saw my dad. By this point, I had read a little bit about people seeing apparitions of loved ones "visiting." In a few remarkable ones, multiple people had seen the same apparition at the same time, sometimes even in different locations.

I didn't think about that immediately. In my dissociated half-asleep state of extreme grief, I just saw my dad. He was there in his thick winter coat and a green and blue plaid scarf he always wore. It was a scarf from Ralph Lauren that I had given him for Christmas/Hanukkah. He never liked the gifts anyone gave and always, right upon opening them, would bluntly say, "You don't mind if I exchange this, do you?" No one got truly mad. Everyone accepted this as one of his many eccentricities. He could never be bothered with pretenses, no matter what. But this scarf, he loved. A real accomplishment.

Wearing it now, right there again in front of me, he was sharp and vivid, much clearer and more real than in a normal dream. He looked worried about me and, just as he'd done hundreds of times before, was trying to figure out what to say or how to help.

For a second, I didn't even react, because it all seemed so normal. But then the reality of it hit me, and I jumped up startled.

He was gone before I had a chance to tell him how much I missed him.

6

I Go To My First Medium And Wtf?!

It was finally time for my first medium reading, the one my mom's friend had suggested, the one who wanted me to wait until it had been a full six months since my loss. I dragged my mom with me on the train to Long Island. I was having so many panic attacks from my loss that I was too scared to ride out alone. My mom grudgingly agreed to ride out with me.

We argued on the ride out. I was mad that she had been dawdling and put us at risk of missing our train and, because of our delay, ruining my chance at possibly getting in touch with dad. She glared ahead in the cab to Penn Station as I yelled at her about being irresponsible. I normally would not have harped on her about it and she normally would have apologized.

We did make the train, but hardly spoke the entire ride out. We were both tense. She was sure I was about to have my hopes shattered. I was scared of the same thing, although our way of saying that was to yell at each other.

We took the train to a small suburban town. After grabbing a bite to eat at a neighborhood sushi place, we walked to the medium's office. My anxiety increased with every step. It was almost unbearable. I wanted this reading over with so I could end this suspense.

We finally arrived at a normal looking little home.

When we rang the door, I expected to see an old woman with a turban, long dress, and tons of jewelry who would lead me into a room with a crystal ball. I reminded myself not to buy a crystal and to not pay for anything other than my session.

Instead of some Madame Fortune Teller, I was greeted by a woman who looked exactly like any other suburban mom. She welcomed me and my mom into her home, which was comfortable and looked completely normal. Photos of her and her family covered shelves and the walls as well as the little knick-knacks you would expect in a family home. While I skeptically took everything in, I could find nothing suspicious or off-putting. There were no crystal balls on display anywhere, or any other psychic clichés.

At first, she seemed a little taken aback that two of us had shown up instead just me, but she invited my mom to stay in the living room, while she and I sat together at her dining room table.

Still super suspicious, I waited for her to ask me my full name so she could feed it to her assistant who, listening in on the hidden mic, would Google me, and feed the medium information about me through a hidden earpiece. Maybe they even used a special background check software only known to a mafia of mediums?

But she never asked my last name.

After a tiny bit of small talk—and, I mean, really small, nothing about my family, job or anything identifying—she asked if I had ever had a reading before.

Me: Only at parties for fun.

I reminded myself to appear respectful and not too leery, but also to pay attention and watch what she was up to.

She told me that if no one I wanted to speak with came through for mediumship, we could just do a psychic reading about me and my life. She further explained that she had no control over who came through. It might not be the person I was there to see. There was nothing she could do about that.

A chill of anticipation shot through me. What would that even mean to have my loved ones visit? Would my dad come through? A part of me really HAD let myself think this could all be real. It was like hoping chemo would heal Stage IV cancer with a two percent survival rate. Your heart hopes, but your head knows better. Not completely impossible, but far from probable.

As I was considering the odds, she seemed to go into what I imagined was a trance, or the mimicking of a trance. She semi-closed her eyes, and her head gently rocked side to side, but not dramatically. It was all pretty subtle.

I held my hands with my palms facing downward an inch above the table. I don't know why I did that. I don't remember if she told me to do that or if I chose to. I then felt this weird warm tingly energy intensify around me. I had never had a feeling exactly like that. It felt like that anticipation energy in my stomach when I was excited about something but outside of my body and around me. And these warm anticipation tingles stayed there. I didn't question them. I was mesmerized and very in the moment.

I waited a few minutes while her head bobbed gently. This weird buzz continued around me, mixed with some intense curiosity and a bunch of anxiety and hope.

Medium: I have a K name here. Who is the K name?

Kenneth. My dad's name was Kenneth.

A chill of astonishment ran through my body. I knew it was just an initial, but still. Was this really happening?

Me: My father.

This was surreal. My throat choked up when I said it. How was it my dad was actually gone? And how was I here talking to a… medium?!

Neither of those things fit into anything to do with life as I had ever understood it to be.

Medium: K—your dad is standing to your right side.

As she said this, a major chill moved into my right side and stayed there. Yes, I KNOW about the power of suggestion, but the experience and the growing sensations were more powerful than logic.

Medium: What happened on March tenth?
Me: I have no idea.
Medium: Did something happen March tenth? A birthday, a day someone passed away?
Me: It honestly doesn't mean a thing to me.

No. I had no idea what she could mean by that date. I felt a sinking wave of sadness and disappointment. Of course, this wasn't real.

Medium: I'm sorry but he won't let it go. He insists it is something.

Was this a tactic of mediums not to admit they were wrong and let the client make the information fit? Despite my disappointment, the physical sensation of energy stayed strongly around me.

Medium: Do you want to go get your mom? Maybe she would know.

I called my mom into the medium's dining room and told her what was going on.

Me: Okay so dad is here.

My mom gave me a look that was mixed with irritation, suspicion, and a twinge of curiosity.

> **Me:** What is March tenth? He is insisting something happened on this date.
> **Medium:** Like his mother's birthday? Or your anniversary?
> **Me:** Yeah… anything?
> **Mom:** No. Nothing I can think of.

My stomach sunk a little further.

> **Medium:** Here's what I'm seeing: a three ten. The number three ten.

A look of amazement took over my mom's face.

> **Mom:** Oh my god! Our building. We are at three-ten East eightieth street.

I felt a huge chill and little bursts of excitement, which pushed away my sinking disappointment.

Of course. Our apartment. Where I grew up. The apartment he lived in. 310—I want to go back to 310, I was just at 310—what he had said daily and often incoherently while in the hospital.

> **Medium:** That is it. He is laughing like he can't believe it was so hard for you to get that.

That is exactly how my dad would have responded.

Looking baffled and dazed, my mom went back to the other room.

> **Medium:** I'm getting he was an older man when he passed. He was in his mid–eighties.
> **Me:** Yes!

Wow. Considering I didn't look like my father would be that old, I had to give a few evidential points for that one, and I felt a slight excitement jolt again into me.

The energy and chills were sitting around me now and not moving away. It was the feeling you get when you hear a beautiful song and chills pour over you. They normally run through you, but it was as if they had poured onto me.

Medium: Your dad mentions you did connect with him. You had a dream and he visited.
Me: Yes.

Okay, probably everyone dreams about their deceased loved ones.

Medium: Now, I have an older woman here. She keeps saying I'm just like her. And pointing to you.
Me: Oh, my grandma. I've been told my whole life we are very alike.

I should have just said, "Yes, I know who that is," but the chills were strong, and I was not expecting correct information after correct information to keep coming in.

My mom was 17 when she lost her mother. I had heard stories of her and how our interests and looks were very similar.

Medium: And she has a small child with her. One who is around two.
Me: Oh my god! Yes.

She had lost a small child. Her daughter died when she was only two. That happened when my mom was a baby, so my mom did not remember, but had known about it.

What in the fuck! This was actually seeming to be real!?! I felt more intense spine tingles but all over my body.

I GO TO MY FIRST MEDIUM AND WTF?!

Medium: Your grandma said you have been living with your mom since your dad's passing. She says it's okay you're doing that for a while, but eventually it will be time to get back to your real life.

How was she doing this?!

Medium: Is your mother a psychiatrist?
Me: Yes!
Medium: And she's incredibly successful? Your dad is showing me a couch and this prescription sign, which means to me she's a psychiatrist. And then he is showing her as the best in the field. One of the most renowned world-famous ones.

If you believe she was honestly talking to my dad, this was incredibly sweet and touching as well as evidential. Yes, as you know my mom is a psychiatrist. And while I think she is great and her patients love her, she is far from world renown. But to my dad (and all of us) she is the best. My dad always would brag about my mom, saying she was the best in her field, and according to him that was the 100 percent truth.

Medium: Your dad wants you to know that your cat, the one who just died is with him. He has your cat.
Me: Wait, what?

How in the fuck?!!! Yes, shortly after the loss of my dad, I also lost my cat. Thanks universe. And go fuck yourself too for that one.

Medium: Yes, your cat who just died. Your dad is taking care of it. He says it's the very sweet and cuddly one.

Yes, this cat was exceptionally sweet and cuddly. My dad would often comment on that and laugh about it. Had I told her my cat had just passed away when I came in? I didn't think I had, but maybe I did. I know I had not posted about it on social media either. Since the loss of my dad, I had disappeared from posting.

Medium: Who is Mandy?
Me: Mandy?? Mandy? I had a friend at school. Actually, she went to a different school. We hung out some as teens, but weren't close.

I thought that, of course, she could have just been saying common names and then was planning to let me fill in who this was while making it look as if she got this information.

Medium: No. It would be more significant than that. That just doesn't feel right.

Hmmm—she did not take the easy route there?

Medium: Oh, wait, do you work with social media?
Me: I do!
Medium: And did you travel to Thailand?
Me: I did! And it was an especially life-changing trip.
Medium: Okay, that is why. My daughter is named Mandy and she also works in social media and is about to go travel Thailand, so that was why I was seeing her in relation to you. So, you can learn a bit about how we get information. It's more like charades than talking on the phone with them.

She continued.

Medium: This is interesting. Your dad is sending me a burst of the color green. It feels like this burst of love.

Green was my dad's favorite color and the color of his office studio room in our apartment. I would often go there to feel safe as a child. We also both always chose green candies when I was little. HOW WAS SHE DOING THIS?! That was not something she could have Googled. Another wave of chills ran over me.

Medium: So, both your grandma and your dad are saying it's

okay for now, but that you're not working. That you need to get back to your work and life.
Me: Yes. True. Can you ask them how? How can I do work or anything when I feel this AWFUL.

I desperately needed to cry to my dad and tell him I felt horrible and ask him what the hell I could do about it. That was another of the hardest things about loss. Often the person who helped and supported you through the hardest things in life was not there to help you through the hardest thing you had ever gone through.

Medium: Just do it. That is what your dad said. He showed me that Nike swoosh. Just do it.

While that was not the gentlest response, that was so completely him. At times, when he was still with us, it could be maddening, but now it was beyond comforting.

I felt a chill slide along my left side.

Medium: Okay your dad is leaving now. Which is kind of sudden. They are not normally so sudden. He went around to your left side and gave you a hug.

That chill built up and lingered on my left side before she said that. This was really so weird!

Medium: He's a nice person. He also stopped to thank me for connecting us. They don't always do that. He's so considerate and thoughtful.

Yes, that would be something he would do.
The medium was quiet for a second, then she laughed a little.

Medium: As your dad was leaving, I asked him if he wanted to say anything to you, such as that he loved you. He responded: "She knows what I think of her."

Hahaha! Wow, that was all so incredibly him! If someone who knew him very well was to create a script of how he would act with a medium that would be it.

I sat there surrounded by the most amazing feeling of tingling chills. I did not want it to end. I did not have any sense of time or how long the reading had been.

Then the session was done.

I was absolutely stunned. I felt an odd combo of shaky and stupefied, yet in another sense calmer. The physical ache in my stomach was lighter and was being replaced with this warm tingly feeling.

Maybe I could come to her weekly? But she explained she would only see clients once every six months. She essentially turned me down as a regular paying customer?!

Then instead of trying to sell me an expensive crystal, as I had been on guard about, she gifted me one.

Medium: Also, be on the lookout for signs from him in the next week. He will send you little signs to let you know he's around.

My mom had come out during this time and heard the final part of our conversation. When we left, she was stunned by the 310 and by the fact that this medium did not want me as a repeat client.

Mom: That goes against anything I would have ever expected.

I filled her in on my session.

How could she have known what she knew? From Googling my phone number?

But that could not have told her about my cat who died, its personality, my dad's personality or how my grandma had lost a two-year-old child. A full background check could not have told any of that aside from maybe the loss of the two-year-old. And that would have been digging way back.

I then realized something else a bit strange.

I GO TO MY FIRST MEDIUM AND WTF?!

Me: Mom, you know what? It was so weird. I held my hands palm down over the table, but hovering about an inch above. I never actually touched the table. They were in the air and it was comfortable. Warm and kind of as if they were lying on this warm buzzing liquid or something.
Mom: Oh my god! When I went in, I realized they were just floating above the table and I noticed, but for some reason I did not even think about it. I've never seen you sit like that and in fact I've never seen anyone sit like that. I can't imagine it being comfortable.

We both mimicked it later back at home. It was not comfortable.

Me: What do you think that was? Energy I was sensing that could be created by the vibrations of strings in string theory? Or could it be the substance that would host our consciousness once we are out of a body? Would it all be stronger around a medium or would I just be connected through her to sense it more?
Mom: I have no idea at all. It's just strange.
Me: Something is going on. It really is. I actually think it is all real.

The wonderment of what I was saying washed over me in more chills. There are no words that can do those chills justice. They were around my body all night. They seeped into my aching stomach and alleviated the worst of my grief pains.

And how had I gotten so lucky with this relatively unknown, low-key medium that she turned out to be so good and (very possibly) genuine?! In one session with a medium, the entire definition of the universe and how I knew it to be was profoundly shaken. Since nothing could better explain what I was feeling, I kept replaying a quote in my mind that I had recently read by Dr. Dean Radin in *Entangled Minds*[1] about Spooky Action.

> *"You'll know you've got it when your gut suddenly drops, like the feeling of free-fall when a roller coaster plunges off that first steep rise. Until you get it viscer-*

ally, the most profound discovery description seems like overkill. Afterwards, profound isn't strong enough."[2]

That night, I slept deeply for the first time in over six months. The next day, once the high wore off, my logical mind began to tear everything down. I did give my real phone number so I could not trust she didn't get information that way. Had I told her my cat just died when I came in? I must have. Had I gotten so swept up in a few astounding moments that were actually just tricks she used to disarm, that I missed the catch—the way magicians distract with misdirection.

Nevertheless, there was still the possibility this had actually happened, and if it had, I had uncovered some clue in the biggest mystery of all of humanity. I needed to figure out what had really happened.

7

Did I Get A Sign?

Although the skepticism and questioning thoughts kept rolling through my head, I HAD seen this medium do something that seemed genuine, albeit impossible. This "laws of the universe defying experience" made me feel good enough that the next day I got out of bed and took a walk. I had not taken a full walk in about six months. I was still disoriented in this world where I was less loved, but I also enjoyed the air, movement, and sunshine.

While walking, I kept thinking over my medium session. What exactly had happened? And she had mentioned I would get a sign. I had heard a little about signs in my research. Dr. Gary Schwartz had touched upon them. It meant that somehow your loved ones would use or manipulate material in this world to show they were around. Apparently finding things such as coins or feathers in unexpected places were signs. One example Dr. Schwartz had mentioned in *An Atheist in Heaven* that he co-authored with Paul Davids, whose friend had passed away. Davids had said to a deceased friend, "Okay, if you are around, I want to be bitten by a spider."[1]

He was.

And then he found out that coincidentally other members of their friend group, as well as Dr. Schwartz, were all bitten by a spider that week. Also, before I had ever heard of the idea of a "sign," Dr. Kleinman had mentioned during the talk at my mom's institute that bird in Jaipur who had shown up after the death of that woman's grandmother. Her grandmother had always loved that kind of bird, and that sophisticated, modern woman assumed this bird was her deceased grandmother paying her a visit.

At the time, I had pretty much thought signs sounded like bullshit. But since so many things were putting a dent in what seemed to be possible versus impossible, I was willing to at least consider.

I was walking down the street in New York and digesting all the astonishing things I had been learning, when I suddenly felt? or was taken over by? this strong awareness to pay attention. I didn't hear a voice or anything, it just came into my head to pay attention to my surroundings. I did without contemplating.

The best way to describe it was that I felt very day-dreamy and at the same time hyper-aware. It was a lot like, for example, when you focus on a painting at a museum and while you are hyper-aware of that painting, you tune out everything else. I looked around and suddenly every person on the street was wearing green—it was not one of the busiest streets like in Times Square, but there were maybe altogether 20 people on this first street and each person was wearing green. Men's shirts and women's dresses. As that medium had said he would do, my dad was sending me a sign. The way he had sent her a burst of green during my reading, his favorite color, he was now sending that "burst" to me.

For the next few blocks, 80 to 90 percent of the people were wearing green.

I also felt this laughing, playful, and warm sense surrounding me, as if there was this energy dancing around me the way you can walk into a special occasion and the air just feels electric. It felt like that, but more personal and much stronger, as if this energy was laughing and playing directly with me.

I kept giggling to myself (hopefully not too obviously) and felt as if this was definitely the sign the medium meant. That my dad was around me enjoying this moment too. I got so swept up in it, I wasn't even questioning it. I then came to a green painted store called "Happy Tails" with an orange painted paw print. Simba, my cat who the medium said was with my dad, was a large orange cat. I then saw a couple a few feet away, who were both wearing something green and sitting against a building with the girl leaning onto the guy. They were obviously very happy and very much in love.

I felt a strong wave of that electric energy and a feeling of love. It was as if I was being shown this couple in love as a demo of love at the end of this experience, which had also led me to a sign of my cat letting me know she was happy, too.

Then, as suddenly as it had come, that electric energy was gone. The second the buzzy, laughing, playful energy lifted, I didn't see an abnormal amount of green clothing during the rest of my walk. My mind was suddenly sharp and clear as if I woke up from a dream.

What the hell was that?!!

Once I was down from the high, it being a "sign" seemed nuts. It must have been a coincidence, and I must have gotten swept into wishful thinking, which felt, of course, exhilarating to believe my dad was around me.

Anyway, how would this have even worked as a sign? My dad somehow had everyone who would be walking down a certain street at a certain time get dressed that morning in a certain color? Or he played with the neurons of my brain, so everyone's outfits looked green? It was okay to entertain any idea, even about signs, but I needed to keep it as experiments only and not take it too seriously. I paid attention during the rest of my walk home (and over the next few days) to see if I tried to notice everything green, would it seem like everyone was wearing green. It didn't. Maybe only one in 20 people instead of 17 in 20 people were wearing green.

When I got home after getting this burst of green from my dad, I felt... What? Happy? The warm tingling of the signs stayed with me for a few hours. Something had happened. I had had a tangible experience that made me think on another level that there was some (possible?) truth to signs. And therefore to survival of consciousness. To my dad still being around me. Evidence and experience kept piling up showing that the impossible might actually be possible. And this went beyond my own grief to the hope that I myself and everyone I loved would never die in the way I had always imagined death. The shock of that, something I could never have fathomed, began to bring a spark (albeit small) of life's vitality back to me.

Then about a month later something else happened.

Boats had always been important in my childhood. My uncle, my dad's brother, had owned one and we would stay on it every summer for a week.

I was taking another walk. It actually felt nice to be outside and moving even if I wasn't ready to throw myself into socializing and working. My stomach hurt less and that heavy molasses-like weight that had seeped into my entire body was starting to lighten up. Ever since the medium reading and my subsequent sign, I had been enjoying this comparative lightness and going out more.

I decided to walk through Central Park, past the Central Park Boat Pond. It was an area I had played in as a little kid. While walking, I felt myself slip into that same day-dreamy state.

I decided, although decided is the wrong word since I gave it no thought, and compelled is too strong a word, so for lack of a better word I was drawn to slow down and start meandering over to the pond. This meandering was odd for me. I am a born and bred New Yorker and no matter how much time I have, I NEVER walk slowly. I find slow walking irritating and trapping, but I felt dreamy and peaceful as I moseyed over to the side of the pond.

The Boat Pond sometimes has a few mini model boats that hobbyists float. I saw one docked along the side of the pond and tranquilly walked over to get a closer look. I cannot explain why I did this. I just did.

As I gazed at the boat in a meditative state, I suddenly noticed it had my dad's name painted on the side of it!?!

Wait... WHAT. THE. FUCK?!

I looked closer. Was this in my head? No. The boat was named Kenneth. I didn't feel startled, though, the way I would expect to be. While logically my mind was screaming, "What in the living fuck?" my body just remained serene as if this was the most normal thing ever.

As I started to think about the unlikeliness of his name on the boat, it was as if I slowly started coming out of a dream state and the wonder of it began to hit me in a slow-building wave.

Then the thoughts came pouring in. How did this happen? How common a name is Kenneth for a boat? Why did I feel motivated to walk over and stare at this in the first place? Could I have seen the name on an unconscious level from far away and therefore felt drawn to the boat? That might have made sense if this was a real boat, but a model boat was so tiny it could barely fit more than one Barbie doll. If I did perceive the name from that far away unconsciously, that would be paranormal itself. I looked a few more times to make sure I actually saw his name on it.

I did.

"I should take a picture," I thought, but then I felt something like a wave of logic wash over me that felt physically similar to disappointment. If I started photographing and getting emotional over every coincidence, how could I keep a clear head and assess my medium readings and the evidence I was studying? So, I walked away feeling a bit disoriented and very excited, but I reminded myself to calm down. There are a lot of coincidences in the world. There are even studies that people highly underestimate how likely it is that something coincidental will occur. [2]

Looking back, if I could do it over, I would have taken a photo. Still, there was enough going on around me, suggesting that the laws of the universe seemed to be different than I had thought. I had to figure this crazy puzzle out!

8

Searching For Psychics And Ghosts

In my quest to crack this "laws of the universe are different than I thought" case, I found a place called the Rhine Research Center. They offered classes with a scientific perspective on paranormal and psi, a word used by parapsychologists for anything that occurs that is inexplicable by normal means. The Rhine was originally started within Duke University, although now they were independent. Their tagline was "Bridging the Gap Between Science and Spirituality."

I could get on board with that.

They had a class on ghost hunting taught by someone called Loyd Auerbach. I assumed this would be different than that 80s movie *Ghostbusters*.

I did some research on Loyd Auerbach. He was a parapsychologist and had written a lot of books on a variety of paranormal topics. He was also a stage magician, which gave him an insight into exactly how paranormal phenomena could be, and often was, faked. When he conducted ghost hunts, he could spot when people were fabricating the phenomena. He seemed skeptical, thoughtful, articulate, and well… normal.

So, I signed up for an 8-week online course on ghost hunting.

Some of the things we learned in this class were pretty worldview changing, if I was going to once again open my mind and at least consider them.

We learned about technology to perceive ghosts, that a ghost could be the consciousness of a passed away person (referred to as a ghost or apparition), or a recording of energy that was no more conscious than watching an old movie with an actor who was now deceased. This was referred to as a haunting.

We examined specific cases and learned that poltergeists, which literally means "angry ghost" in German, are usually caused mentally by a living person and their energy. The intense emotions this living person felt created an energy so powerful that it actually moved objects, but the person was unaware they were doing so. They were often as baffled as anyone by these moving objects. This moving of objects with the mind was called psychokinesis, or PK for short.

Parapsychologists used a technology called EMF (electromagnetic field) meters, to see if there is any correlation between any unusual electromagnetic fields in the environment and someone, such as the ghost hunter or a medium, reporting experiencing or perceiving a ghost. Electronic Voice Phenomena (EVP) devices (actually, anything that records sound, even your own phone) can be used to pick up any supposedly ghostly sounds.

Sometimes ones that no one heard at the time of the investigation, but then would appear during playback. People could ask questions such as what is your name, and see if a voice would be imprinted and play back when they listened. A medium would also often join, and Loyd Auerbach explained that having a human there to assess was more important than meters. The meters could react to things such as a refrigerator. Or, one time, people were living on a highly radioactive area, which was medically dangerous, but they assumed the headaches and reactive meters were due to ghosts. Luckily, the ghost hunters got there in time to let them know it was not ghosts!

Loyd Auerbach explained that TV shows and media sensationalize what ghost hunting is—not the biggest shocker there. The real stuff was actually a lot more interesting than the TV sensations, but a lot less dramatic. One of my favorite things I learned in this class was that our senses are much more culturally based and subjective than we realize. For example, a Pygmy tribe in Africa had always lived amongst trees and, therefore, these people had not developed a sense of depth perception. When they were taken to an open field, they perceived buffalo in the distance as bugs. As the buffalo moved closer, the tribe believed the bugs were getting larger.

The way we perceive and interpret the world, whether you are a skeptic, a religious fundamentalist, or a tribesman living amongst trees, is how we are taught and how we think the world should be. The way it actually is, is much less important to our perception than we realize. My perception of the potential of ghosts shifted from one of stories at summer camp and horror films to one worthy of scientific investigation and research.

The one thing that left me uneasy was that the weird things happening outside the laws of science did not promise that there was consciousness after death. Objects that flew, for example, could just be from living people's minds influencing them, and seeing ghosts could just be imprints, like a tape recording of previous people with no consciousness. Paradigm-shifting, yes, but not a promise that I would see my dad again.

It also dawned on me that if parapsychologists were using this technology to detect if there were ghosts, and that medium I went to sensed that there were a few presences around me—actually not just sensed but gave some startlingly concrete evidence for—wouldn't that mean that there were "ghosts" around me? Loyd was careful to make sure we understood a change in EMF did not mean there was a ghost. No technology exists that can prove there is a ghost. However, if there were ghosts around wouldn't they always cause an impact?

Suddenly, every flickering light or overhead footsteps in my parents' apartment became a possibility of a ghost of my loved ones. How could this be scary to people? Ghosts, if real, would just be people—us in 100 years, our loved ones now and in the future. I needed to see if there was a ghost. Or, more accurately, I needed to see if these energies that medium told me about checked out at another level with this ghost-hunting technology.

I sent an email to Loyd Auerbach asking for a suggestion of a ghost hunter. Loyd replied as if this was a very typical email to send, which in his world it was. He gave me the contact information of an actual ghost hunter called Dan Sturges.

I reached out.

I got a form email back from Dan that explained not to panic about whatever phenomena was going on and that he wanted to set up a phone call with me before we met. I was glad about having a call first since I assumed there was a higher probability than average that anyone who called themselves a ghost hunter was a bit nuts.

When the phone rang at our scheduled time, I laughed at the absurdity of the fact I had a call with an actual ghostbuster. Then I felt a knee-jerk-reaction of guilt for laughing. How could I laugh when my dad was no longer here?

Dan was friendly and seemed surprisingly grounded. He asked me to tell him what was going on. I told him about my recent loss, the fact that I was taking Loyd Auerbach's class, that I had gone to a medium, that, yes, I knew how completely weird that was, and that the medium said I had a few of my loved ones around me. If they were around me, wouldn't those read on ghost-hunting equipment?

He wanted to know if I had seen weird phenomena.

I let him know that lights seemed to flicker a lot, but that I had no idea if they did so before. I had never paid attention.

> **Dan:** I have to let you know, almost every time it turns out to be nothing, so I don't want you to be disappointed.
> **Me:** I won't be. I don't even believe it that much.
> **Dan:** The next step is to meet in person since if we go forward with this I would be coming into your home.

Ah right! He wanted to be careful before coming into my home because he probably thought there was a good chance that I was crazy.

Me: Of course. I assume you get a lot of crazies if they think there are ghosts around.
Dan: Some have mental problems, which can be heartbreaking. Some are lonely. With some of them, it's nothing more than they have been raised to believe in ghosts, but it turns out to be just some normal thing like footsteps from a neighbor upstairs. Some are really curious. And sometimes inexplicable things have happened.
Me: Ah okay. I am not crazy or anything. Well, I kind of think maybe I am a little bit since I went to a medium and am talking about ghosts in my home and all. I mean that is what crazy people do and talk about. Right?
Dan: You don't sound crazy. And I'm sorry for your loss.
Me: Thank you. Oh, I also need to ask, how much will this all cost?
Dan: Nothing. I consider it research. Scientific research. And I appreciate your letting me use your home to do my research.

Really?! So maybe most people in this world were not just preying on naive and grieving people for money?

On the day I went to meet Dan, I put on a pair of jeans, a tank top, and a little lipstick in a semi-attempt to show I was a human. Oh, and I even brushed my hair. And, yes, I did shower.

My cousin, my dad's niece, was over, lounging on the couch. A year younger than me, she had grown up a few blocks away from me and was more like a sister. We were close, but being in her first year of grad school for psychiatry, she was not only a lot like my mom, she was even more skeptical than her.

Cousin: Wait. Are you wearing clothes? Not pajamas?
Me: Yes.

Yes, she was teasing me, but she had been a total rock since the loss of my dad. Despite her teasing, she never tried to correct how I was coping with my grief. Something anyone who has had a significant loss knows is rare. I also knew she missed him a lot too.

Cousin: Are you leaving the house?
Me: Yes. I have a meeting with a ghost hunter.
Cousin: Hahaha… So, for real, what's bringing you out of the house finally?
Me: I just told you.
Cousin: Oh. Oh my god! You are serious. Umm–okay. Well… lemme know how it goes.

I took a seat in an outdoor cafe by Bryant Park, where Dan and I planned to meet. When he showed up, he was not what I was expecting. He was a completely normal, friendly looking guy. I was expecting someone who looked like they lived in their parents' basement. A knot (one of many) slightly loosened in my stomach.

Maybe, if normal people did this stuff, there honestly could be something to it?

We began talking. I noticed that he was also skeptical. When he went to haunted houses for ghost hunts with mediums, he would meet them at a spot and then take them to the location so they could not Google anything about the site. Some found that annoying. Regardless, he stuck to keeping everything evidential.

He explained he got into this because he loved science and energy. We talked about experiences on his ghost hunts and even string theory.

He mentioned there were two well-known scientists: one who had attended a haunted house investigation and another who had a long talk with him and agreed that with scientists' understanding of energy, ghosts were possible. That was encouraging! They had asked him to not reveal their names because it could hurt their careers.

Maybe in secret they weren't so dismissive of the paranormal? I had always idealized science as having nothing interfere with the truth. I guess nothing done by humans can run as perfectly as it should in theory.

He warned me to be very careful because there were tons of classes promising to teach about psi, but many were cons. Not shocking.

Me: But how is the Rhine?
Dan: The Rhine is very credible. Congrats on finding the best one. Here is an example of a con type—I once took a class on astral projecting (going out of body) and I went to the intro and the guy was going to charge everyone three hundred dollars to learn how to astral project. He claimed he could project to other planets. I had my ghost hunting equipment and said if he astral projected right now and it registered on my equipment, I would pay for everyone's class. He kicked me out.
Me: Oh yeah! I don't trust that. If I thought I could astral project I would WANT to test it. I would also go spy on the Windbridge mediums to see if they were just Googling. And I would spy on how they prepped for their tests with Doctor Beischel. By the way, do you know her? Doctor Beischel?
Dan: I do. I have met her a few times. She is very smart.
Me: Really!? Is she? (another knot slightly undone) So... umm you think there is actually something to all this?
Dan: I do. But not with all mediums. And also no one can know for sure what this ability they have is, or where it's coming from. My favorite mediums say they have no idea how or why they can do this. They don't assume it's from dead people. They say they don't know.
Me: But you think there's a chance it's true that we survive bodily death? That it really could be coming from dead people?

I tried to keep my desperation out of my voice. I wanted him to feel free to be completely honest.

Dan: Yes. There is a chance. But I can't say I know that. It's just one possible explanation of what's going on.
Me: Oh—uh. Of course, that makes sense. I had one reading with a medium so far. She isn't from Windbridge or anything. She was actually surprisingly good. At least I think she was? I felt great afterwards but I realized I gave her my real phone number so I can't prove she didn't just Google me.
Dan: She could have.

Me: Do you think she did? She got a part of my address but didn't know it was my address. She kept asking what does March tenth mean—three-ten. I kept trying to figure out what is March tenth and suddenly it was like, "oh my god, of course." And she got my dad's initial of his first name.
Dan: If she Googled, she probably would have gone for gold and given the full name.

Wave of relief!

Dan: But the only way you can be positive is if you hadn't given your number.

Wave of sickness. I couldn't believe I had been so careless.

Dan: Give a Google Voice next time.
Me: That's what I was thinking. I'm so mad I didn't do that.
Dan: You also should check out Forever Family Foundation. The founders lost their daughter, and I think they had some not great experiences with mediums, so they decided to test and then certify mediums and weed out the bad ones. Email them and say you are going to mediums and using precautions such as Google Voice. They would be interested.
Me: I have heard of them. I'll have to look them up and learn more. But one thing confuses me a lot. So, this medium said these people are around me. And most people, when they go to mediums get the same thing. But then when you do ghost hunts most of the time nothing shows up on your equipment. Wouldn't these energies show up?
Dan: Good question. I don't know why.

We finished up and said our goodbyes.

Me: It was great to meet you.
Dan: You too. We can figure out a time to do the test at your mom's place.

Me: Parents' place. Great. Thanks. Oh, and one more thing—you really did meet Doctor Beischel, right? And she really was smart?
Dan: YES. And I'm sorry again for your loss.

I reached out a few times to arrange a ghost hunt, but it never happened. I think Dan clearly saw there was not any abnormal phenomena going on.

A year later, I emailed and asked Loyd Auerbach the same question about why someone could have a deceased loved one show up in a reading with a medium, but not on a ghost hunt. He emailed back with the following reply:

To: Loyd
From: Liz
Subject: Re: Ghost Hunting?

Hi Liz,
First, mediums are ostensibly communicating with people 'on The Other Side,' rather than ghosts who have not gone to 'The Other Side.' Mediums make a distinction between ghosts (apparitions who have not transitioned) and spirits (entities on 'The Other Side').

Second, there has been little done with any environmental sensing equipment in/around a medium when doing a typical communication session. Environmental sensors have been disturbed when a medium or psychic is communicating with an apparition in a location (a ghost).

Finally, it's not clear that any of the tech actually reacts to ghosts. The actuality may be that it is intentionally influenced by apparitions (aka ghosts). They (the tech) may, however, react to something correlated to haunting experiences/place memory (the ones that are like recordings with no consciousness. Or even PK, living people's own minds).
Loyd

Once I completed my ghostbuster degree, I continued my studies at The Rhine, taking a four-week class called, "How to Find a Psychic" and another four-week class, "How to Develop Psychic Abilities." According to Loyd Auerbach, there are three kinds of psychics (able to read minds or energy of living people) and/or mediums (able to connect with those who have passed) to watch out for: the fake mediums who know they do not have abilities and con people for a lot of money. They will prey on the grieving, never giving very evidential information, but still hook those in deep grief in by claiming things such as their loved one needs to talk to them one more time. The ones that do not have abilities, but someone once told them they were psychic, and they now think they are but are only just somewhat intuitive. They do not mean to be deceptive even though they are.

Then there were the worst kind—mediums or psychics who do have abilities and they use these abilities to trick people and get a lot of money out of them.

Among the worst of the intentionally fraudulent are the fake psychics and fortune-tellers who will initially charge a low rate, but then state you have a curse, which they can remove with crystals, candles, a spell which all cost significantly more than their initial fee.

Entertainment psychics you would see at a magic show, also apparently get real information at times. It had happened to Loyd Auerbach and it had happened to some of his friends. If it could happen to them, could it happen to regular people like me?! What would that even feel like?

He reassured us that despite all the phonies, there were genuine mediums who were also ethical. Phew. I had had a few more readings since my first, but the second one had seemed like she meant well, but her reading was not very evidential. The third was a total con. A trendy LA medium who had some celebrity clients. I had a phone reading with her. She got nothing right and tried to tell me my dad (who I had to tell her had passed) wanted me to be more trusting. Kind of convenient for her right? She kept asking me what I wanted out of the reading, and my request of "valid evidence that consciousness can survive death" was not anything she could provide.

Loyd Auerbach taught us personality traits of what he thought all good mediums had: a sense of humor and humbleness. They did not take themselves too seriously and they were curious about their abilities and where they came from.

He then brought up the thing that Dr. Julie Beischel had hit up against and researched as well: Are the mediums communicating with our loved ones who survive or just psychically reading our minds?

That hit me like a brick in my stomach. How could I know, even if I got the best reading ever, that they weren't just reading my mind?

I posted in the question section for the class and explained how I had used a hidden identity and all I had done to keep my readings evidential. Maybe Loyd Auerbach would see a hole in my methods? I added that I now felt stuck on how to prove that the mediums weren't reading me psychically. Was there a way I could test for this?

He replied that there is no guarantee, but that some mediums had gotten information that the sitter getting the reading did not know at the time, but then later verified through other family members. A 100 percent pure reading could not exist because mediums were still people and bits of their own thoughts would get mixed in.

He mentioned Dr. Beischel's studies where a proxy sitter got the reading from the medium for someone else's deceased loved one. The medium, therefore, was not reading the proxy sitter psychically because the medium did not bring in information about the proxy sitter's loved one. The proxy sitter did not even know anything about the person's deceased loved ones, so they could test for this. Plus, when I read Dr. Beischel's books, she asked the mediums this exact question. They all said the two types of information felt different. Although that was just their perception, nothing concrete.

I then added in a live chat question section during class, "Maybe it is best to send someone in my place as a proxy?"

Loyd replied, "If you are going to go into a medium reading and not take in any of it and be so suspicious you trust nothing, it is best to not even go in the first place."

From behind the screen, I wrote rationally worded questions. I probably came off as logical, skeptical, and maybe even challenging.

What he could not see was that I attended these online classes wrapped in a blanket wearing the same sweatpants and hoodie I had been wearing for an entire week. My face was grey and sunken, and my eyes were swollen from insomnia and crying. This "skepticism" was coming not from a place of trying to expose or challenge, but from a place of desperation.

I did not know how to appropriately express this to him, so I did not reply. Also, if I revealed that level of desperation, I risked getting a "comfort" answer instead of an honest one.

He did talk further about the differences between psychics and mediums—a line he wanted to make sure we never forgot was that —"all mediums are psychic, but not all psychics are mediums."

He would not teach us that if he had not concluded there were definitely real mediums communicating with our loved ones, not just us psychically?

Right?

WHEN HIS NEXT CLASS, *How to Become Psychic* began, he opened up with a line to encourage all of us who thought psychic abilities were only available to an elite few: "Anyone CAN play basketball, but not everyone is Michael Jordan."

Apparently, psychic senses are just another sense anyone can develop. In our culture, where we are told they don't exist, we don't often develop them.

Loyd Auerbach used an example of The Himba tribe who never had a word for the color blue. When tribe members were asked to distinguish between green and blue squares or different shades of blue, they couldn't. They could not recognize blue existed because they were never taught blue existed.[1]

I had no idea what to make of that!

Maybe not being shown that "this is blue" makes the brain cells that can distinguish the differences between green and blue die? Or maybe those brain cells never activate? And what about those certain people who can see a color spectrum the majority of us can't?

How much of our world are we not perceiving because we were never told it existed?

To increase our psychic abilities, we learned exercises such as one by a well-known psychic-medium who had passed away called Alex Tanous. Since psychic was a type of sense, the idea was that we would increase our psychic abilities by strengthening our other senses: Lie still for five minutes and focus intensely on one sense. Try different senses on different days. Listen carefully and hear every sound including background noise, look at every minute detail, and when eating, taste every texture and subtle flavor.

It was relaxing and intriguing to notice details that I had always overlooked, although nothing happened that I would classify as paranormal. I became aware how we (at least I) only process a tiny percent of what is right in front of us. How does our brain choose what to take in and what to leave out?

Loyd Auerbach told us to get a partner and he would give us a few exercises to try. Crap! I knew my cousin and mom would not be open to this and I was not ready to reach out to my friend group with all of this weirdness. So, I posted on the class forum and a woman reached out. We scheduled our call for the first exercise.

Me: I really don't have abilities.
Her: Me either.

She lived out in the Midwest and was a physical therapist. She did not go into why she wanted to develop these abilities beyond her personal curiosity. I didn't push.

We were to get on a call with our partner and during the call take a few bites of different foods and not tell our partner what we were eating. We were to eat slowly and mindfully, taste every bite. The other person was to try to sense what we were eating, not just guess.

If your partner was eating an apple, you might get a sweet and slightly tangy taste on your tongue. You might then feel the texture of an apple but not get the word "apple." You could feel it was a little cool and hard as opposed to soft and salty like peanut butter. It would feel juicier than bread. That was how psychic senses work rather than labeling and defining with words such as apple.

It was interesting to see how opposite our culture is to being in touch with psychic abilities and intuition. We label, define and know, not sense.

I ate first putting my phone on mute so I wouldn't give away chewing or sipping sounds. The sensation of sweet or salty, or the taste of kale or bread, was different when I noticed every little detail of it. The meaning and whole treat of eating something sweet, for example, lost its power of craving. The food became an experience with multiple flavors and layers and nothing more. It was neither good nor bad nor delicious. It just was what it was.

She began to guess my foods and did not get anything accurate. Then we switched. I closed my eyes and tried to see what I felt. I tried to imagine what connecting with her would feel like, but I couldn't feel another person was there the way you often do when someone is in the room with you. I tried to feel if any sensations of taste came onto my tongue; none did. Many different foods and drinks ran through my mind, but no different than if I was playing a guessing game. Apparently, neither of us was very psychic.

One of the questions that I (and everyone else) want to know came up: if they really have abilities, why can't psychics predict the lottery?

> **Loyd Auerbach:** Think of them like a meteorologist. They read patterns in the current situations and the atmosphere and then can predict probabilities. But it is not definite and unexpected things can change weather patterns. Any psychic who claims they can accurately predict the future is full of it.

I later had this explained further in one of Dr. Claude Swanson's book, *The Synchronized Universe*.[2] Dr. Swanson has become one of my favorite scientists. Let's say a life is represented by a ball rolling down a hill. There were times when there was one path. Times when the ball could veer off into one of two or more paths. Sometimes the ball could equally end up veering onto any of those paths, while sometimes it was much more likely it would veer down one over another or multiple others. Mediums/Psychics could get a bird's eye view and see options, paths, and probabilities. A psychic could see, for example, that all the lottery numbers would be even, or they could get a portion of them, but not all. Or maybe they got all but had no idea what the numbers were for. Or for which lottery and on what day.

That would be so annoying!

And even worse. In terms of predicting disasters such as plane crashes, a psychic might know there will be a plane crash in say, the 3rd week of October, but they won't know what day, or time, or airline. What can they do? Call every airline and say they are a psychic and ground all planes on said date? Even if the information got more specific, such as a city, the city wouldn't shut down an entire airport for a week over a psychic prediction.

Another knot loosened in my stomach. This was the first logical explanation I had heard for why psychics aren't billionaires stopping every disaster.

As I slowly started to re-engage a little more in the world, I continued to mull over all this "any of us can have some level of psychic abilities" stuff, and I kept returning to an odd experience I had when I was around three years old. I had never talked about it and in fact, until my dad became dangerously sick, I had pretty much forgotten about it.

I had been lying in bed after my parents had tucked me in, read me a bedtime story, and left me to sleep. Then I heard this voice. I felt it attached to a presence, too, although at three years old, I wasn't analyzing it like that. It was just there. I don't remember word for word what it said, but it was a woman's voice that said something like "I love you." Then it said something about how I was special. I partly heard it and partly just understood what it was communicating, which was love the way my parents loved me.

I didn't question it because at that age I hadn't yet learned the laws of the universe and what "can" happen versus "can't." It was just another new thing in this world where things I didn't yet understand happened daily. I did become curious, as I was about all the new things I kept discovering.

So, I went to my mom.

Me: Mommy, I am hearing these voices talking to me in bed.

I remember my mom taking a deep breath, looking at me thoughtfully and saying:

Mom: Are they mine and Daddy's voices?
Me: No.
Mom: Or the television? Is it the neighbors or people outside?
Me: No. I dunno?

While we did live in an apartment, you couldn't ever hear the neighbors, but you could often hear voices carry from the street. However, I knew it wasn't from the street. It felt different than when I heard those voices. It felt closer to me physically. It carried an emotion, and it was there for me and cared about me. It was a strong presence, the way I felt the presence of my parents when they were in the room.

I sensed (but I didn't have the vocabulary for it at the time) that I was alone with this, that this was something my mom didn't understand, and my dad wouldn't understand. That was frustrating because I wanted to understand it. It was also frightening that there was something my parents couldn't explain to me and I couldn't communicate to them.

I never heard the voice again and I never brought it up again. I still thought about it maybe once every few years. When I did, I was curious and a little frustrated. I logically assumed it must have been our neighbors or voices from outside or a half-asleep dream.

But a part of me always viscerally KNEW it was very different than that.

The first time this memory had reemerged was when I was

starting to realize (if that is even possible to realize) my dad was not going to get better. I was on the subway, reading one of the first books that opened up this new world, and the possibility of time travel, *Time Travel in Einstein's Universe*.[3] Suddenly, that memory came back clearly to me. I somehow "knew" it was very important and tied into the book I was reading and what was happening with my dad.

I can't say I logically put that knowledge into left-brain words, but in some deep knowing way I felt its relevance to now and what I was to do next.

Although I thought it was something like "I love you," I tried to remember exactly what this voice had actually said, but I couldn't.

Forward to a night about six months after my dad had passed away and I was immersed in all this research. I came down with a piercing headache. I rarely got headaches and this one was brutal. I took a Tylenol and fell asleep.

Early in the morning, I woke up to someone saying my name.

Liz.

They said it once, very strongly and crisply. It was a woman's voice. I looked around, but no one was there. I assumed I imagined it. Then I heard it again. Strongly.

Liz!

I definitely heard it!

And then I heard it a third time.

This time, I got up to check.

The only person in the apartment was my mom. I checked in on her, and she was in bed, sound asleep. She never talks in her sleep. Even if she had, her room wasn't near enough for me to have heard.

I realized I recognized the voice. It was the same woman's voice from when I was three! It was warm, but matter-of-fact, clearly pronouncing each syllable. I felt her presence and I felt an excitement coming from her knowing that I heard her.

The experience of hearing this voice also felt different from how I normally heard.

When I heard normally, there was a slight vibration and processing of sound through my ears. My ears could tell how close or how far the source of the sound was and whether it was on my left, right, in front, or behind.

However, when I heard my name now, the voice was just there in my brain. It was a very subtle difference, but I still noticed it. I realized that was also how I experienced that voice was when I was little!

It then hit me, I HAD experienced "hearing" like this a few years back in the real world. I was at a product fair. A friend was demoing headphones for deaf people that stimulate the part of your brain that hears sound, while bypassing your ears. I took the headphones, which were flat circles that I was told to place onto my head itself. I plugged my ears with my fingers and still heard the music playing in the headphones. It was like hearing clearly, while feeling nothing in either ear and a slight movement in my brain.

Did this mean there was some way this presence directly reached my brain? And how did I know to tie in that memory right when my dad got sick?!

9

Weirdness At A Medium Workshop

It was 3 a.m. and I was scrolling through my emails (grief insomnia!) when I saw one in my "fake email for mediums" inbox saying that that Windbridge Medium, Laura Lynne Jackson, was holding some event called "Connecting with The Other Side" in New York.

I had now had a few more mediumship readings, including with a medium who was certified by Windbridge, all of varying degrees of accuracy. It had been two months since my first one, and eight months since my dad had passed and when I first learned about Dr. Jim Tucker, but this all still seemed REALLY unbelievable. Not as in holy shit unbelievable, but as in there was some catch that explained it that I still hadn't discovered.

I debated going. Aside from the occasional walk or coffee with a close friend, I hated leaving bed, but I decided, "If I'm gonna do this, I'm gonna go all in." I clicked through the prompts in my phone and bought a ticket and then FUCK! I realized too late it was connected to PayPal and therefore to my own name and credit card.

How could I have been so stupid?

I debated canceling and getting a friend's card and rebuying the tickets, but I partially didn't have the energy and partially figured if she now told me Googleable info, I would have a pretty good idea of what was going on. Also, just because she had my name, she wouldn't know it was me when she saw me. Ugh, aside from the fact my face would show up in a search.

After going back and forth, I decided that I would go to the event and I could see what information she gave to me. Then I would have to move my reading with her to the phone, so she would never put two and two together, the way she could if I was sitting in front of her. In her book, she said she preferred phone sessions anyway, so that gave me the perfect excuse.

When the day of the event came, I dragged myself out of bed. Despite the glimmers of hope from my research, my grief was still intense and I barely engaged with the outside world. My stomach hurt all the time. My legs hurt and felt heavy as I struggled to get dressed and get moving the way I had in my prior life less than a year before.

I wondered again, did I even want to go? I pretty much knew what was going to happen during the event anyway. Laura was a great presenter and would be very dazzling and spew a bunch of crap that she knew was crap such as our loved ones, well, loved us. Or she was honest, but a total new-ager who would tell me in an overly sweet voice stuff like the universe loves me and that I need to have trust.

Maybe she had amazing connections somewhere like the CIA or an uncle in the FBI? Or she had an amazing private detective she worked with. Maybe that type of connection was what separated the best mediums from the mediocre. She would come out with accurate facts such as "wait is there a Sara Golden here? Your grandfather Michael Golden from Miami who died on November twelfth, two thousand and twelve says hi and he loves you."

And then there was the other option that I wanted more than anything. That she was for real. But to entertain too much hope for something like that was terrifying. Because with a science-minded background, you expect that if you get far enough along, you will eventually discover the twist in evidence that exposes none of it as real. All built-up hope shattered. How would you ever recover?

Mom: Honey, I think you should probably shower and wash your hair.
Me: Why? Everyone there won't care. They are probably those weirdos who don't use deodorant. Anyway, I don't care what Laura thinks. I'll never see her again. I switched my session to phone.

But I knew I could not sit in a room full of people after not having washed my hair for a week. So, I showered, washed my hair, but left off makeup.

I dragged myself into the elevator, managed a smile at the doorman, and then I was outside. The cheerful contrast was a brutal reminder. I was halfway to becoming an orphan.

I was late, which was super embarrassing. I had assumed it wouldn't start on time since it would appeal to flaky people, but the room was already packed and Laura had started speaking. So, I did the one thing I did not want to do as I climbed over people in rows of folding chairs: draw attention to myself.

I sat down. Once I was sure Laura wasn't aware of me, I felt under the seat for a microphone or some recording device. I had heard that mediums recorded your conversations with the other audience members then later fed you back the information you gave. I had also heard mediums hired plants to befriend you and ask why you were there and details about who you lost, which they then told the medium. I was not going to share anything personal with the other guests.

I balanced listening with an open mind with evaluating everything she said—and didn't say—for the catch.

And... well maybe? She could be for real?

Dr. Beischel had tested her and the tests and controls sounded pretty rigorous. It seemed that however hard I racked my brain, there was truly no way someone could pass this test by luck or fraud.

Right?

But could Dr. Beischel just be in on it too?

These thoughts spun in my brain, and my stomach reacted to every single one.

As I listened, I noticed that Laura seemed really likable. Not likable in that charismatic leader way that gurus tend to be, but sociable and fun. Like you wanted to be her friend. In fact, she actually reminded me of some of my friends. And not that weird friend either. She was a vibrant blonde in a stylish, but not flashy dress, who reminded me of the kind of girl I hung out with in college doing too many shots and meeting cute guys.

She didn't preach. She didn't try to coerce anyone into believing her. She explained things incorporating science and brain scans and mentioned she liked evidence and hated things that were all new-agey. She never even gave us any readings or acted as if she knew anything about anyone—and I assumed a lot of us had used our own credit card.

She explained that we are all connected to those who have crossed over to The Other Side and how these discarnate consciousnesses—our loved ones—are still aware of us. She saw cords of light connecting us to them and to people who were still living as she wrote in her book.

Despite the astronomical claims, her self-presentation (combined of course with the Dr. Julie Beischel stamp of approval) made me not dismiss this as absurd. She explained in a very practical and relatable way how her brain worked, mentioning the EEG brain scans she had gotten by Dr. Jeff Tarrant and her participation in scientific studies. For example, she would see a screen in her mind during readings. Her psychic information would come in on the left of the screen and mediumship on the right of the screen. Dr. Tarrant's brain-scans backed that up. The left of her brain lit up during psychic and the right during mediumship.

She came off as a balanced and smart person who somehow randomly was born with this ability our society deemed delusional. And she justified it scientifically, logically, and confidently.

One woman in the audience raised her hand. She began to speak, and her voice shook. I remember nothing of the question, but I do remember exactly what she looked like, what she wore, and I also remember the first line she said: "I lost my daughter… " The woman was fairly young. I assumed her daughter was a child.

I had met a few people who had lost a child before, but I was protected from the reality of it by the distance of being a kid. This was the first time I was in a peer group of people who had experienced such a significant loss.

It hit me that if I was going to dive into this world, I was not only diving into a scientifically fascinating world, but one filled with other real people like me who were suffering from tremendous loss.

Laura handled it kindly and, for the first time, I saw her as someone who must deal with people living with this level of agony every day. What was that like being around people in that level of pain all the time? I wasn't sure if I was intimidated by the magnitude of realizing what this new world consisted of, or relieved that I was in a room of other people who felt as shitty as I did.

Laura told us about making signs with loved ones. Unlike how my dad had (hypothetically) surprised me with the green and the boat, we needed to choose our signs and, when we ask our loved ones to show us they are around, we would see that sign. But if you focus on seeing something, anything, don't you always just notice it more? This is called the Baader-Meinhof phenomenon, a frequency illusion.

She continued, explaining that deer, birds and butterflies appearing around us could be signs from our loved ones too. I guess like that story from India Dr. Kleinman shared of the grandmother's favorite bird appearing?

I raised my hand.

Me: But... how? I don't get it. They... become a bird?
Laura: No. It's not like that. They can affect objects because they are energy. For example, birds and deer navigate using electromagnetic fields. So, our loved ones can use their energy to affect the EMFs and direct the bird. They can affect metals and electronics too.

I had so many more questions, such as what form of energy were they? We can't see air, for example, but we know it is a molecule called oxygen made up of atoms. And it can combine with two hydrogen molecules to create water. So WHAT substance is our consciousness? I shot my hand up again, but she had turned to answer another person's question.

And this explanation of signs? Well, that actually had some logic to it. At least more than anything else I had heard.

Instead of calling on me again to start chipping away at my long list of questions, she told us that now we were going to make a list of our own signs we wanted to get from our loved ones.

I had nothing to lose. Plus, I had had my sense of what was possible so drastically challenged recently that I was not going to rule anything out.

I wrote a few things for my dad: poker, which he played all the time; a song he used to sing to me about a cat in a box; and, finally, a green stuffed rabbit he and I used to play with when I was little and made up a whole personality around. Worth a try. Right?

We then moved onto another exercise, which involved remote viewing printed out images that were enclosed in an envelope, then drawing what we thought these images were.

I had learned a lot about remote viewing in my research. I had also been reading about remote viewing experts—Ingo Swann, Paul Smith, Harold E. Puthoff and Russell Targ. They had even participated in remote viewing experiments called The Star Gate Project conducted by the CIA. Apparently, this group had remote viewed some shockingly accurate details of classified locations. The details of The Star Gate Project were declassified in 1995. [1]

I had also recently learned about a fascinating group called the Society for Psychical Research (SPR) that conducted a lot of scientific research on remote viewing as well as on mediums. They were still around with headquarters in New York and London. I was planning to see if I could volunteer for them, but the only issue was it seemed like all the studies I had found were conducted in the 1800's–1900's.

I closed my eyes, tried to calm my mind and imagine what was in the envelopes, but nothing came to me. So, I just drew any random thing I could think of. I was not even close but—then there was a kind of, I was sure meaningless, coincidence. One image was of a cat in a box. That was the song I had just made as a sign with my dad. But, why out of the other 60 or so students in the class, who all wanted a sign as much as I did, would that drawing be a sign for ME?!

As the class continued, I remained mesmerized by Laura's "normalness" despite her claims, that this consciousness living out there — with no brain—can (hypothetically) communicate with all of us?! And it communicated clearly and accurately with her.

Yet the way she explained it—electromagnetic fields, energy, etc.,—kind of made sense. It was no more out there than the multidimensional universe stated by string theory. Or how about "shadow people" which would be dismissed as pure sci-fi, was it not being claimed by Stephen Hawking. Yes, Stephen Hawking claimed that "shadow people" from other dimensions were reflecting from a different dimension onto ours. It is definitely worth listening to the YouTube video where he explains this![2]

I raised my hand.

Me: But, if life after death is true, why do the smartest people, like Stephen Hawking, say that it can't be true?
Laura: He's just one person and not all scientists say it's impossible. Einstein saw a possibility. And Stephen Hawking hasn't dedicated himself to studying this.

Laura's response wasn't mind-blowing, but it still offered a little bit to think about.

I immediately Googled "Einstein thoughts on life after death" and I found that he both believed and did not believe in an afterlife. Ugh, of course! This world must be obscure no matter what!

The "do not believe" argument was in response to religion, not any parapsychological studies:

> "Neither can I believe that the individual survives the death of his body, although feeble souls harbor such thoughts through fear or ridiculous egotisms."[3]

The "believe" argument:

> "Now he (his friend Michele Besso) has departed from this strange world a little ahead of me. That means nothing. People like us, who believe in physics, know that the distinction between past, present, and future is only a stubbornly persistent illusion."[4]

Einstein said this many years after his "do not believe" quote, so it appears as though he might have changed his mind? Even if he didn't directly say that he thought there was an afterlife, it did seem he was describing time and consciousness as more of an illusion than the "consciousness is a result of the brain" theory.

I raised my hand again.

> **Me:** But what substance is this consciousness?
> **Laura:** It's energy.

Hmm... so if energy cannot be created or destroyed, as Einstein said, and we are energy... what did that then mean for us?

> **Me:** But literally. Is there a molecular structure? Or substance. That this form of energy is made up of?
> **Laura:** No. It's not this measurable thing. It's an energy. The energy of love. I do, however, think physics will end up proving this all one day.

I wanted to be able to measure it, though. In a lab. Run by scientists.

But wasn't there a time people knew they breathed in and out. And felt wind. Way before we knew "oxygen"? Maybe we just weren't there yet.

We then broke for lunch and I decided to introduce myself to Laura and ask another question.

I waited in line for my turn, debating if I should even introduce myself with my first name since it was on my email to her. When she gave me my reading, she might then know it was me. But Liz was a common enough name. Right?

Me: Ummm hi. I wanted to meet you because well...

Was she annoyed that I had been sitting there looking super suspicious and had asked skeptical questions?
But she was actually incredibly nice.

Laura: Oh my god. I love you. You have the best, most positive energy. It's so great to meet you.
Me: Really?! Me? All I do is lie in bed and cry, but thanks. And... uh... I also have a question. To be honest, I umm... well... ugh... I hope I don't offend you, but I don't fully believe any of this. But I have been reading all of Doctor Beischel's studies and books, so I am giving it a try. Anyway, when I book sessions, I give a fake name. Will that ruin my readings? Because in Julie Beischel's studies they give a real first name. So, could a fake name make a medium connect with the wrong person or be unable to connect or anything?

Laura wasn't offended. She told me it was a great idea, and it was important to be skeptical and questioning of all of this. I should keep doing that.
I then told her I was also taking classes at The Rhine.

Laura: Oh, I LOVE them. I do stuff with them a lot.

REALLY? She did? Hmmm—so Loyd Auerbach respects her then?

Laura: Well, it was nice to meet you. You feel familiar. I feel like I'll be getting to know you.
Me: Uh, maybe? Nice to meet you too.

I knew she could have been so nice to me to disarm me, but that was just one possibility, not the only possibility.
I continued to lurk nearby, pretending to be checking my phone as I subtly tried to listen and assess what she talked about and with

whom. I was taken aback to hear how many people came up to her crying and then told her who they lost. Why would they give away a key piece of evidence? Maybe they were plants? Or just believey.

As the line of questions ended, I watched who she went to lunch with, to see if those same people would try to befriend me or anyone else in the room and ask us personal questions. I surveyed the area. Was there anyone putting secret recorders anywhere, or leaving "forgotten" phones, or whatever the catch was?

Looking back, I am not sure I was as subtle as I thought I was.

After the room cleared out and I checked around a bit (but came up with nothing), I went downstairs to the cafe to eat lunch and absorb everything.

When we returned to the classroom, Laura was standing there with her friend, who was wearing head to toe green, my dad's color. I didn't wanna get superstitious or anything, but it couldn't hurt to at least note this observation.

Before class picked up again, I ran to the bathroom. Her friend happened to also be walking out to the bathroom.

> **Friend:** Hey. I want you to know I heard what you were saying and the questions you asked in class. I have seen so much stuff at this point. It really is true.
> **Me:** You honestly think it is? Like what? What have you seen?
> **Friend:** So much. I think, if you keep exploring this, you will discover some remarkable things.
> **Me:** So, honestly. You truly believe this?

I felt my throat tighten and I started tearing up.

> **Friend:** I do. I promise.

She asked me for no details about my life or loss.

I returned to class and Laura handed out dice. She had us pick a number between one and six, then roll the die twelve times. The odds of our chosen number coming up more than twice was higher than chance. This would mean, in theory, that our minds were affecting matter— through psychokinesis. Like the Random Number Generator and poltergeists.

She told us that if we got significantly under chance, such as zero, that could be our mind affecting the dice to not let us succeed, if this went too much out of our realm of what we considered possible. I later learned this was called the Sheep-Goat Effect.

I picked my lucky number five, took a deep breath, thought of it and rolled my die twelve times, but nothing significant happened. I think I got my number three times. When we were done, Laura asked who got their number twelve times? No one. When she got to ten, there were a few people, and a high number of people seemed to get it nine, eight, and seven times.

What were the odds against chance for a certain amount of people in a group this size to get highly beyond the odds of chance? It did seem as if an abnormally high number of people got their number at a significantly higher rate than chance.

While contemplating these thoughts, something kind of weird happened. I felt this intense heat, but not burning or painful or anything, shoot from the die into my hand and up the lower part of my arm. At that exact second, I vividly remembered a Windbridge medium reading I had, where he was shown dice by my dad. I felt as if there was almost a flash in my brain as I made the association that this was what he meant. At the time, I said I had no idea why he was seeing dice.

I then pulled myself back to rationality. I remembered that brains have the innate need to make patterns. What must have happened was I remembered what that medium had said, my hand sent the signal to my brain, and my brain, having a need to make this more than a coincidence out of the need to believe this stuff, sent a hot energy feeling to my hand and a "holy shit!" signal to my brain.

Laura then asked us who had abilities. I knew by abilities she meant paranormal ones. Quite a few people raised their hands. Obviously, I did not.

Laura: Who here often has this feeling of hot tingling energy in their hands?

That was pretty clearly a manipulative question to make us all think we have abilities. EVERYONE has that feeling in their hands! Hands are extremities and, therefore, there is extra blood flow, which causes warm intense tingling in palms and fingers.

I raised my hand with a slight eye-roll and looked around wondering who was going to think this made them (and everyone else who happened to be here) special. Huh? I was one of three people to raise their hands. In a room of about 60. Wait, doesn't everyone feel that?? Wasn't that the nature of… having hands?

Laura: Great! That means you have strong natural abilities to interact with and feel energy.

We moved on despite the fact that I was a bit weirded out by apparently having a different experience of having hands than the average person.

After guiding us on a 15-minute meditation, Laura told us to pick a partner. I turned to a friendly looking woman with long brown hair and glasses. She looked about my age.

Laura: We are now going to give each other psychic readings.

Wait! What?
I turned to the girl I was going to "read."

Me: Sorry, I have NO abilities. I don't even believe this is all real.

She smiled a little shyly and said she had no abilities either, but did believe it was real. I agreed to give her a reading first, since not believing this I was perfectly okay with the fact I would get nothing for her. I turned to her and I began to guess a bunch of drivel. The first thing that popped into my head was "Grandmother," but I decided to try to guess something a little less obvious.

Me: Umm uncle.
Her: Yes.

Okay, many people have lost an uncle.

Me: He had brown hair.

That was an easy shot since this girl had brown hair.

Her: Yes.
Me: And three kids—two girls and a boy?
Her: Oh my god! It was THAT uncle. I thought at first it was a different one.
Me: Wait... what? REALLY! I'm just making this up.
Her: YES! This means so much.
Me: And there was a house that was central to the family that like all of you would go to every summer or something?
Her: YES!!
Me: And he was on your dad's side, but older? And there were a lot of kids in that family? Like five or six when he and your dad were growing up.
Her: Yes!!!

Her eyes were filling with tears. SHIT! I was making this all up. I didn't want to trick anyone.

Me: And did his name begin with like D or L?
Her: No... But those were two of his kids' initials.
Me: SERIOUSLY!?

Tears were now running down her cheeks. Fuck! I felt so sleazy that she thought this was real. I couldn't let myself deceive her.

Me: Ummm... You do know I just made it all up. It was—uh—it was a coincidence.

But every single thing I had guessed was right.
And she kept tearing and thanking me!
She still believed I had truly talked to her uncle even if I didn't believe I had.
Nevertheless, that was BIZARRE! I was too stupefied, and she was too emotional for us to switch.

Laura: Okay how did that go? Did anyone get anything?

A few people shared—yes, they had.
I then raised my hand.

Laura: Do you have a question?
Me: Uh not exactly—but uh—this is strange, but I got a lot right. I just started guessing and saying a bunch of stuff and I got like a million facts right. Not literally one million, but I guessed six things and they were all right. And I pushed myself past the obvious grandmother guess. This is really WEIRD!

I noticed that my hands were shaking.

Laura: Ha. Interesting you choose the word guessing.
Me: Well, it WAS guessing. I just made it up. Honestly.

Everyone in this room made up mainly of people who believed in "defy-the-laws-of-how-scientists-say-brains-work" stuff seemed mildly amused by my insistence that this was just lucky guessing.

That was the last exercise of the day, and I was still completely shocked as we began to gather our stuff. As the class filed out, I lurked around to see what people said to her. I also wanted to see if the last exercise was just a way for her friends to pretend to read us while actually gathering information on us to share with Laura. No one seemed to interact with her in any way other than regular students. But, of course, if she was good, they would not give it away so easily. They would email later. Or call to avoid a digital paper trail.

But half the class being plants to get five minutes' worth of information on people, many who were not necessarily even on her waitlist, did not seem that likely either. Her readings apparently lasted two hours, so these five minutes would not cover a full two hours of information. Also, I had followed her Facebook page. She rarely had events and frequently gave readings, so she had significantly more one-on-one sitters than people in her classes.

Plus, I wanted to see, given any of this was true, who WAS this person who actually talked to dead people? And Dr. Julie Beischel agreed that she did. Also, if this WAS true, why did this seemingly ordinary person have this incredible ability I would give ANYTHING to have. If only I could talk to my dad the way she must be able to talk to her loved ones.

After I pushed my lurking to the limit, I walked home, processing this emotional and confusing day. I didn't know what to think. I had not found the catch. It didn't mean I had concluded there was no catch, but I was no closer to having any insight into what it even could be.

I also REALLY liked Laura. Of course, I did not know her well enough to form a valid opinion of her, certainly not one close to worthy of challenging the very laws of the universe as I knew them to be, but I had none of that funny feeling that probably all of us get when someone doesn't feel genuine. My gut could not be a deciding factor that would override science, but I took it seriously enough that I gave this all another small level of credibility.

I concluded it might be time to try to do something besides just lying in bed. I was planning to contact the ASPR (American Society for Psychical Research) to volunteer, but I wanted to know what other suggestions she might have. I had to explore this further, so I decided to send Laura an email.

To: *Laura Lynne Jackson*
From: *Liz*
Subject: *Nice to meet you!*
Hi Lauren*,
Thanks so much for an amazing class. I was wondering if you knew of

any labs or places like Windbridge but in New York or LA. I would love to volunteer for one that does those kinds of studies.
To refresh your memory, I was the sciencey-skeptical one, but still nice and open to evidence.
If you have any suggestions, I would really appreciate it.
Thank you.
Liz

```
*Yes, I got her name wrong, but I didn't
notice this until later.
```

She replied.

To: Liz
From: Laura Lynne Jackson
Subject: *Re:* Nice to meet you!
Hi Liz.
I am so glad you came to this class and gave it a try. That must be fun to live in both cities.
You should try Forever Family Foundation.
Good luck.
Laura

Right. I kept hearing about them.

When I got home, I checked out the Forever Family Foundation website. It said you needed to be a member for six months before volunteering. Membership was free, so they weren't using that to make money. I signed up by filling out a form and made a note to myself to contact them in six months.

What would I uncover if I could get a behind-the-scenes view? Would I find the catch? Or would I discover a world beyond my wildest sci-fi, matrix-ey, multi-dimensional dreams? And maybe even find where my dad was?

10

I Finally Figure Out What I Am Just Not Seeing

I had been feeling… better? Not the way I was before, but even if I still didn't want to go to parties or events, I was meeting friends for coffee. I was taking regular walks. I still ached for my dad, but the possibility that an afterlife could actually exist brought a daily excitement and purpose to my life, an excitement I once had for my career. I kept trying to return to my career and get my startup, which was about to launch an e-commerce site, up and running, but I could not muster the zest I once had for it. It was as if I had gotten all these clues, that what I had assumed was a fantastical world actually existed. And that was so much more interesting than any startup.

However, the idea of an afterlife, a dimension where our consciousness would continue, still did seem as if it could only be fantastical or fantasy. How and where and in what form could our consciousness survive? When I thought of it that way, it did not seem possible. Still, I had seen so much evidence, concrete evidence, so what was I missing?

You know when you go to a magic show and are mesmerized by the performance? You can contemplate for hours attempting to figure out how they did the trick. You know that there is something you aren't seeing, something that would be obvious to a well-trained magician.

So, what was the catch I kept missing?

I could not find the catch in Dr. Jim Tucker and Dr. Ian Stevenson's research, nor could I with Dr. Beischel's studies of mediums. There was Loyd Auerbach's classes at the Rhine, Laura's workshop, the research and studies I had poured over, plus my own medium readings.

The deeper I delved, the more hidden the "normal" explanation seemed to become.

I read every skeptics' comments and articles I could find, yet none of their alternative explanations or arguments were very strong. I kept digging. All my life I had been taught that skeptics were the logical ones grounded in reality. The ones undigging and uncovering cultural fantasies that power- hungry leaders used to oppress people and charlatans used to con people.

However, most comments and articles that I found were angry and snarky. Snide instead of logical.

Another typical approach by skeptics was not snark so much as disregard, a dismissive eye-roll instead of an insult.

The third approach was the most reasonable. They did come up with arguments that were intelligent. However, their arguments were always addressed in the books I had been reading and controlled for in the experiments—these skeptics never mentioned that.

While there were many valid uncoverings of con artists, no one had validly uncovered the Windbridge Mediums, Dr. Beischel, Dr. Tucker, or those of that caliber. It honestly seemed as if no skeptic had delved very deeply into the topics they had such strong opinions on.

But that did not mean there were not clear-headed skeptics who would see the "normal means" explanation, that I, untrained, had missed.

I decided to reach out to a skeptic.

I found one with a more thoughtful approach than just "this all wasn't true because it's not," while offering no counter explanations. He said on his blog that from all of his experience and research he had no reason to think any of this was true. Everything he had studied, seen, and experienced ended up having a "normal" explanation, but he had studied it all in-depth and he was still open. If anyone could show otherwise by offering evidence that this was true, please go ahead.

I emailed him and made sure to clarify I was not the kind of person who normally would ever see or believe a psychic. I felt embarrassed sending a message I would have before considered absurd to the kind of person who would very probably dismiss me as absurd.

But I sent it anyway:

To: Skeptic
From: Liz
Subject: Question
Hello.
I am sure you get many crazy emails, so I hope this isn't too insane. I recently discovered your site. Due to a personal loss, I have begun to explore the possibility of psychics and mediums (something I would have never believed before). I am trying to keep an open mind and exploring if there is any possibility in the reality of this world.

I am sure you have a million things to do but if this would be of any scientific/experimental interest to you, I would love to share how I am contacting/paying for each to see if there are any holes you can find in my approach. I would love to share the recordings of my readings too and see if any holes can be poked as well.

To contact mediums—I use a fake email address, a VPN, and Firefox —the one browser I normally never use in case any of my info could be stored, a Google Voice number, and I have a friend (not a family member or anyone I work with) pay. If it's Zoom or Skype, I use a fake account and I won't add them until
just before my session, in case they are able to trace etc.

I FINALLY FIGURE OUT WHAT I AM JUST NOT SEEING

I have had a few so far—one really impressed me, but I have to eliminate it bc I gave my real phone, one had a few things surprisingly right, but most wrong and a bit foggy, one purely could be applied to anyone and nothing factual, and a few others blew me away and I really cannot figure out the catch.

Would love to see if you could figure it out!

Thanks!

I knew his reply was probably going to shatter the hope I had built up so far, but if I did not seriously examine it, I could not trust my assessments. I was also incredibly curious, even if I knew it would shatter me, to see how the mediums had pulled off what they had.

Two days later, he got back to me.

Before opening his reply, I gave myself a few minutes. This was probably my last few minutes I could have a hope of seeing my dad again. And of not being obliterated myself one day. Then… I took a breath and opened it.

To: Liz
From: Skeptic
Subject: Re: Question
Thanks for your email. Look my best advice to you is to save your money. Don't give these frauds money, even if under the idea that you can expose them or poke holes in what they do.

There are so many ways to con someone.
Take away the chance to cheat and all mediums fail, 100% of the time, no exceptions. So that leaves cold reading, and unless you are an expert in these techniques, you will be conned.

You are welcome to publish your results on my forum. But the way you are doing things is wrong, you are giving them money. I would be more interested if you were trying to get refunds.

I have had over 250 readings with mediums and I have never paid. THAT is more important.

Hatred jumped out of my phone.

I felt a surprising wave of protectiveness for the mediums I had met and the scientists whose work I had read.

I was disappointed by his complete lack of curiosity and questioning of something he supposedly dedicated a good part of his life to. Despite what he claimed on his site, he did not seem at all interested in "anyone who could show otherwise and offer evidence that this was true."

He showed no interest at all in my results, who I was, or how I came to the conclusion I did. He didn't even address what I was saying; he only wanted to talk about money. If they ALL cheat, how did he explain Dr. Julie Beischel's studies? Literally, how? I was not being rhetorical.

I ran to show it to one of the most skeptical people I knew, my mom, to see if she agreed.

Me: Ugh. That skeptic got back to me. I learned NOTHING.

I ranted as I took her through each line.

Me: "Look, my best advice to you is to save your money." I hadn't even brought up money! My one and only question was that I wanted to know what protection I could have missed when I booked or went to a reading.
Me: And then—"Don't give these frauds money, even if under the idea that you can expose them or poke holes in what they do."
The only "holes" I said that I wanted to "poke" were from a scientific curiosity. Not shaming them. I thought he was a skeptic? Wouldn't he want to help me figure out the puzzle? I did a TON of his research and data gathering work for him.
Me: And then—"There are so many ways to con someone." I KNOW! What though! What are the ways he means and how had I not protected against those?

I FINALLY FIGURE OUT WHAT I AM JUST NOT SEEING

Me: Also—"Take away the chance to cheat and all mediums fail, one hundred percent of the time, no exceptions." I explained how I HAD done that. Or at least how I thought I had. How had I not done that?

My mom was listening quietly as I continued.

Me: "So that leaves cold reading, and unless you are an expert in these techniques, you will be conned." Isn't that exactly what I HAD protected against? And exactly what Doctor Beischel and Forever Family had protected against? I had hoped to send him my recordings and have him let me know how I was being conned or cold read.

As I had already explained to my mom, cold readings are when fake mediums give a general reading based on how a person looks—age, class, race, etc. and then further notices and plays off the sitters' responses. It is just people reading, and some are very good at it. Mediums cannot do that on the phone with someone they have never met, as many of my readings had been. Also, some of the information they gave would not have been possible to get that way. Think about my first reading—my cat, the color green, my grandma losing a baby. And other mediums had given me names.

Me: How is the fact that he didn't pay for his readings the most important aspect of his research? I wanna know how he protected his identity and why has he come to the conclusion he came to? If he put this much time into it, wouldn't he love to share his reasoning and results?

And then it happened. One of the catches I just hadn't seen clicked. These vocal skeptics who I always idealized as curiously and unbiasedly examining evidence were just as closed-minded as any true believer.

What would happen if these skeptics and more scientists dug as deeply as I was starting to? As other logical-minded skeptics such as Dr. Beischel and Dr. Dean Radin and William James had?

11

Spoons, Psychics And WTF?!

Me: I am going to this psychic medium workshop out in Long Island. Wanna come?

My friend Jerome was in NYC from LA. I doubted he expected to join me at a psychic event.

Jerome: REALLY?
Me: I know that sounds weird, but I went to a workshop of hers before and, honestly, she's this completely normal person. Her name is Laura Lynne Jackson. And there's this woman Doctor Julie Beischel who was a pharmacologist…

I filled him in. He agreed to come. Jerome had grown up in Guam in a culture that was open to this. He also believed in God, so he wasn't the skeptic I was. In fact, his sister had gone to a medium after their father passed, but that medium was more concerned with selling her a set of expensive candles than offering evidential messages from The Other Side.

Me: I promise this one isn't like that.

We walked from the train. It felt good to move again.

The event was held in a town called Hauppauge in Long Island. It was summer, so it was nice and warm, and the air felt clean and smelled of grass. It felt like an energizing escape from the city. I felt the same nervous anticipation I had grown accustomed to since delving into this research. Of course, I hoped this would add more concrete evidence to the probability of an afterlife, but it could also shatter this newfound hope. While that sounded dramatic, it was not so different than many other days so far. I also felt a little more alive, knowing I was going to be seeing other humans, who were also in grief and interested in an afterlife.

When we finally reached the Radisson Hotel, Jerome pointed to the entrance.

Jerome: Umm... I assume we follow those people.

We entered the massive lobby, which was decorated with typical tacky hotel chain carpeting. But the normalcy of a standard American hotel felt, well, normal. Like a bit of memory from my previous life before I lost my dad—an odd merging of my previous normalcy (well, relative normalcy at least) with this new crazy world, where the laws of the universe no longer seemed to apply.

Large groups of people were checking in at a booth in the lobby. And they were all dressed in costumes. One guy was decked head to toe in white, and I recognized a number of Star War characters. A pack of guys were dressed up as Ghostbusters. A green ghost, I believe called "Slimer" from the movie, danced around them.

Jerome: Umm... I guess these are her fans? Or clients or something?
Me: Yeah... Uh... I guess so? She works on ghost hunts so maybe...
Jerome: You went to another workshop of hers. Right?
Me: Yes.
Jerome: Was everyone wearing costumes?
Me: No. No, they weren't.

Jerome: Well, what exactly were they wearing?
Me: Clothes. Normal clothes. Jeans, tank tops, one woman was wearing this cute dress…

We followed a robot to the check-in booth.
I approached the attendant and tried to avoid eye contact.

Me: Hi. Do we check-in here for Laura Lynne Jackson?
Check-In Girl: Who?

Oh God, don't make me say the psychic event!

Me: Laura Lynne Jackson. The one who does research on consciousness?
Check-In Girl:… ??
Me: She gives talks about the latest scientific research and data on discarnate consciousness.
Check-In Girl:…???
Jerome: The psychic medium event.
Check-In Girl: Oh, no. This check-in is for the Cosplay Convention. The psychic medium event is over there.

She directed us to a room, where a small crowd—thankfully dressed mainly in jeans, sweaters, and other regular clothes—was filing in. I gave my name at the door, still kicking myself for having fucked that up in the first place. I made sure Jerome paid cash and signed in without giving his real name.

I saw Laura. She was animatedly talking to a few people. I noticed she was wearing a stylish dress and very fashionable heels. Her makeup looked perfect too. I looked down at my sneakers and yoga pants. I don't think I had even unpacked my makeup from when I hurriedly moved home as soon as I heard my dad was sick. It had been a while since I had put on my designer clothes and shoes I used to love, did my makeup and went into the world.

I watched who Laura was speaking to and how she was directing the conversation. Was she asking leading questions to get people to reveal anything? I didn't detect anything suspicious. We took our place in the last row. Again, I subtly felt under my seat for a microphone or any kind of recording device. I told Jerome to do the same. We didn't find any.

The horrible ache of my loss was still intense in my stomach, but I noticed that surrounding that awful feeling was the stirring of a thin layer of happiness and hope. I listened to Laura explain how mediums work. She said that during readings the wave patterns of mediums' brains match the wave patterns of brains during sleep. Brain activity in their frontal lobes turns off. She also told us that most of us are actually born with psychic and mediumship abilities, particularly when we're young. But at around the age of five, we enter into our left-brain schooling, when we are taught what is considered possible versus what's impossible. That's when we start to lose those abilities.

Like that tribe's perception of buffalo as bugs!

We learned about how energy is central to everything. This is how it's possible to communicate with The Other Side. We tap into the energy of our deceased loved ones, which still exists, though in a different form than a body. For example, think about physical attraction. You actually feel a physical, palpable energy of another person. We can experience something similar with out-of-body beings.

Before we broke for lunch, I asked a question that had started to concern me.

Me: So, what if our loved ones who have died reincarnate while we are still here? Are they lost to us?
Laura: I know this will sound a bit out there, but I will explain what I was taught. Time and space are not limited there the way they are here, so someone on The Other Side can be with you and with someone else at the same time. Time there isn't linear. I was told to picture a maypole where each ribbon is a single life and all the lives are happening at once from a central pole.
Me: That makes no sense, though. Because, if time isn't linear, how do we gain knowledge and grow? Isn't personal growth

based on cause and effect, which is dependent on time? And how come if we are experiencing all these lives, not in a sequence of time, am I not aware of that? I am only aware of the past of this life and now.

Laura: I know it's confusing. I think as a person it's impossible to understand these concepts. I wish I could understand it myself, but I hope it helps to know that from what I do know, your people are around you, and reincarnation, or anything like that cannot take them away from you.

I was frustrated I didn't get a clearer answer, but I was impressed that she did not try to pretend to understand. Part of what she said was in line with a lot of what I was learning. Time did not seem to exist exclusively in the way we perceive it. And understanding other dimensions like the 4D tesseract, or the concept of infinity, was impossible for us humans in the third dimension, despite our best efforts to study them. I guess I could not expect to go explore the very nature of consciousness and whether we survive bodily death and get neat-and-clean answers.

After lunch, Laura told us we would be doing spoon bending. Okay, that was kind of weird. Just last night I had been reading assigned articles for my classes at The Rhine with Loyd Auerbach. One had been on spoon bending—something I had always assumed was nothing more than stage magic. While reading, I had the thought that if spoon bending proved to be real, that would significantly increase my trust in this crazy world. Probably a coincidence. But still… ?

Laura explained how spoon-bending works, at least in theory. It is actually a form of psychokinesis, or PK, the mind's ability to affect matter. Metal has bonds and grain boundaries, which energy can manipulate.

I listened. I did not know what to think.

Laura passed out spoons. She told us to look them over, make sure they were regular spoons. It was a typical spoon, the same kind I'd use to eat yogurt or ice cream. She told us to try to bend it with our hands. No matter how hard I tried, I couldn't do it; neither could Jerome. Laura guided us to stare at our spoons, tap into our energy and to all yell, in unison, "bend." I got over the silliness of that and went with it, not expecting very much.

I looked over at the woman one row over from me. She had bent hers up into a twisted ball!?!

Me: WHAT THE FUCK?! HOLY SHIT!!

I said that much louder than I intended. A few people looked over at me, laughing at my shock.

Me: Sorry, but did you guys see that?! Look at her spoon!

The woman who bent, balled up, and twisted the spoon laughed and shrugged—as if the fact she just balled up and twisted a metal spoon was something she did regularly.

I took a deep breath and tried to see if I could feel energy in my hands, but so far all I could feel was a cool metal spoon. I took in another breath and focused harder, pushing away any distraction. I felt a warm tingling in my hands. That tingling built up and then what happened next was SO WEIRD! The spoon started to get incrementally hotter and hotter as I held it in my hands, but it didn't burn or hurt. Then I felt the spoon turn soft as if it was melting. Almost turning into a liquid.

I easily bent the head of the spoon down.

Then it cooled to a normal temperature and I couldn't bend it anymore. Some people who were completely unfazed by all of this were contorting their spoons into crazy shapes and rolling them into balls. I tried to think through any logical way this could be happening. Maybe she hired a group of people who knew stage magic to make it look like they bent these spoons? I knew bending spoons was a reasonably simple magic trick, but according to Loyd Auerbach that was a mistake skeptics and paranormal investigators often made. Just because something can be faked does not mean it is faked every time. If you followed that logic, then nothing was real because everything in our lives was faked on TV.

A LOT of people bent them. Did she hire the majority of the audience to impress... well, me? And my friend? And maybe about 15 others out of 40 or so people?

And that still wouldn't explain how I bent it!

Laura told us about an experiment that came close to explaining what I was witnessing first-hand. A metallurgist analyzed metals that were bent normally compared with metals bent "paranormally." There was a difference between the two. Loyd Auerbach wrote about this experiment in his book *Mind Over Matter*:[1] "The boundaries in the structure where the grains of metal touched were fractured and torn in the normal (bent with strength by weightlifters) sample, yet somehow melted or even vaporized in the paranormal sample."[2]

My hands were shaking from the shock. I had melted a metal spoon! With my bare hands. If my own energy and consciousness could interact with and influence metals, maybe—just maybe—my energy and consciousness were also strong enough to exist and affect things beyond our material world. This was also a very tangible undeniable demonstration that what is defined as scientifically impossible (bending metal with my hands!), can actually, at times, be very possible.

I photographed my spoon and texted it to my mom and cousin.

> ME
> Hey. I bent a spoon!

> COUSIN
> Holy fucking shit!

Then from my mom an hour later:

> MOM
> Jesus Christ!!

To this day, the thing my mom finds the most amazing in all of my research is the fact I bent a spoon.

Our next activity was using Zener Cards, a set of five cards with one of five different shapes on them. These cards are used in labs to test ESP abilities.

One of the most mind-blowing studies I had learned about in my classes at The Rhine was called the Pearce-Pratt study.[3] JB Rhine, the founder of the Institute, tested two guys, Hubert Pearce and Joseph Gaither Pratt. One man looked at an individual card while the other tried to "guess" which card they were holding. The experiment included many variations of this. One variation included a man writing the cards they were intuiting on a piece of paper, while the other man was in a different room. To Rhine's surprise, one of them would consistently get a score way beyond chance. In fact, many of that man's scores were close to perfect. They did this both in real time and at different times, where one would "guess" what card the other man had pulled from the deck during an earlier session. These studies demonstrated that, not only did some people possess inexplicable psychic abilities, but their abilities were not affected by distance or time.

And now, similar to Rhine's experiment, Laura picked up one card at a time from the deck. The cards featured five different illustrated shapes, from a star to a squiggly line. We were supposed to guess—although Laura would tell me not to use "guess"—which shape appeared on each card she drew. We marked which shape we intuited? or sensed? on a paper handout. For each draw, we had a one in five chance of intuiting it correctly. We repeated this exercise twelve times.

I took a breath, tried to clear my head. I noticed for each card Laura pulled, two things came to me. The first was a shape that popped into my mind instantly. The second soon followed: a shape that arose in my thoughts more naturally, as if I had pictured it. While the first shape was more just suddenly and instantly there, the second shape slowly took form as I relaxed and focused. To test which shape—the first or the second—was more accurate, I noted both on the handout for all 12 cards Laura drew.

Laura read the results. I got the first one right—it matched the immediate guess, uh I mean "intuit" that had popped into my head instantly. She read the second one... same thing. Third... same thing.

Okay. This was starting to get a little weird.

She went through the rest, and I ended up getting nine out of twelve right with my instantaneous, first guess intuit. Of the other remaining three, I got one right with the second, more slowly developed thought. I got the last two cards completely wrong.

Nine out of twelve right in the "popping into my head"?! That was like Pearce-Pratt level?! I got the highest amount of anyone in the class.

WHAT IN THE ACTUAL FUCK?!

I was still absolutely shocked when we took a break before the mediumship part began. Those who only signed up for the first part left. The crowd became smaller and more intimate. I was eager to watch Laura give a group reading, something I had not yet seen any medium attempt. Whether she chose me for a reading or not, whatever I was about to experience over the next two hours would be emotional and consequential. It had the potential to either help ease my grief and how much I missed my dad, as well as soothe the existential dread of my own impending death, or shatter the newfound hope that was slowly beginning to heal the sharpest layer of my grief. Because Laura was so strongly recommended by Dr. Julie Beischel and Dr. Diane Powell, I was terrified I would lose my trust in them if I uncovered the catch or witnessed her outright cheating or using manipulative tactics like cold reading or giving general information. I would then be left to face the likelihood that my dad really WAS gone for good, my mom would one day be gone, and I would one day be gone too.

As I waited for Laura to begin, my hands started to shake again, and butterflies danced in my stomach.

A desperate energy permeated the room. I had never experienced an energy like this before. (I was now making a point to identify and feel energies.) I badly wanted to be chosen. I desperately wanted to talk to my dad again. We used to talk daily. Not being able to enjoy that call left me with a constant emptiness. He was always full of life and warmth, caring in a way I loved and missed dearly. His first question was always about my pets. "How are the babies?"

I needed this reading. But so did everyone else there. Everyone that is except for one person.

Jerome turned to me, whispering.

Jerome: Ugh, I REALLY hope she doesn't pick me.
Me: What?! Why?

I knew he hated to be the center of attention, but still to not want a reading, evidence that your loved ones are with you, and a chance to talk to them?

Laura stood at the front of the room and looked around at everyone.

Laura: I feel like I am getting pulled over here.

Then she started to come right over to me!

My heart started pounding harder and harder. My hands started shaking more intensely. Was it me? Yes! She WAS coming right to me!

I could feel all eyes move over to me with what I assumed was a combination of jealousy (we ALL want to talk to our loved ones), happiness for me (we GET grief, we are in "the club" and like to see other people have experiences that make them feel better), and anticipation (they hoped she was accurate). The more accurate she was, the more that demonstrated that their loved ones were around too. On the other hand, if this medium, highly admired by the scientists, was not accurate, what would that mean?

And I wanted to talk to my dad!

Laura: Hi, I have someone coming in for you.

I took a deep breath and, despite my shaking hands and pounding heart, reminded myself to sit still and say nothing but "Yes," "No," "Maybe," or "Could you please explain further?"

Laura stood right next to me. I gazed up at her in gratitude and desperation.

Laura: Oh sorry. Not you. Your friend. Next to you.

Seriously? Jerome?!
The ONE PERSON IN THE ENTIRE ROOM who did NOT want a reading.

Laura: I feel a father figure coming in for you.

Looking really uncomfortable, Jerome shifted in his chair.

Jerome: Oh, uh yeah.

Laura started giving a bunch of identifying features. She said that Jerome had a sister and three older brothers, who he wasn't close to. She said his father always did this thing with his tongue that made a certain noise. She knew how he died and how long ago and she mentioned a few other things that Jerome confirmed to be true.

I tried to mentally note each point and check it off to see if she could have known it by normal means and how much she got accurate versus neutral versus wrong; but I couldn't focus enough to do this properly and get a percentage score because I could not stop cracking up at the complete absurdity of the moment.

How was it that the one person who did not want a reading got the very first one?!

I held my hand over my mouth to stifle my laughter. I told myself how serious this was and how disrespectful I was being, but nothing helped. I couldn't stop thinking about how batshit this all was—the fact I was even here in the first place at a medium workshop. At a medium reading. Desperate to talk to my dead dad. Trying to gather more evidence that he wasn't actually so dead, in an air-conditioned banquet hall in this chain hotel. And then of all the people Laura was reading was my friend, Jerome. Jerome! A last-minute invite. Someone who believed in God and had NOT spent the last eight months delving into evidence of an afterlife. And researching Laura. And researching the people who had researched Laura. And most of all he was the ONLY person in the entire room who was not desperately craving a reading. In fact, he didn't even want one at all!

I was so caught up in the moment of the humor and the pure irony, it took me a second to realize how incredibly good it felt to actually laugh again! Which made me happy. Which made me laugh harder.

Is this grief? Just when I had forgotten what the fun of life felt like, it comes back full force. At an inappropriate moment. Laughing like this made me remember how good it felt and how much I missed being a fun person. But what if losing myself in laughter like this would only happen during special occasions now? That fear didn't stay too long because the laughter took over at the complete absurdity of the moment.

I had hardly laughed in close to a year. Anytime I let myself laugh, a wave of guilt and sadness washed over me. But, apparently, all it took for me to laugh again was my friend's deceased father coming in through a medium in a room full of grieving people, during a time I absolutely should not have been laughing.

Laura continued reading for Jerome. Oblivious (at least seemingly) to my inappropriate laughter.

> **Laura:** You don't even want to be here, do you? Your dad is telling me you were brought here.
> **Jerome:** Uh, yeah. Unlike her, I do believe in it and all. It's just kind of intimidating.
> **Laura:** Yeah, and they are laughing about it and knew you didn't want to be called on.

At least I had some company in my inappropriate laughter!

> **Jerome:** Uh. Yeah.

Laura seemed to be a genuinely kind and caring person. In her book, she wrote that those on The Other Side only come over with messages of love and she has never felt anything other than love during readings, even from discarnates who were terrible people throughout their lives.

This one reading seemed to be a bit of an exception. Jerome's dad started yelling at him through her, which didn't help my inappropriate laughing: "WHY DON'T YOU EVER CHANGE YOUR SHOES. YOU WEAR THE SAME PAIR ALL THE TIME. YOU EAT TERRIBLY! YOU AREN'T BEING HEALTHY AND YOU ARE NOT HONORING OUR CULTURE. YOU NEVER EAT OUR CULTURAL CUISINE."

Laura stopped for a second.

Laura: I'm so sorry. I just... I just have to say what they tell me! KEEP STICKING WITH YOUR FRIEND. (she pointed to me—HA!) SHE IS A CATALYST FOR GETTING YOU TO DO THINGS. Uhh okay... umm... I'm sorry but he just said: STOP BEING LAZY AND GET TO WORK. YOU SHOULD BE MAKING MILLIONS OF DOLLARS BY NOW.

With that, the reading ended. I was still laughing.

Jerome turned to me and said, "Uh, well, that's kind of why I hoped not to get a reading. But, WOW, she was accurate! And that is so my dad to push ahead of everyone and go first."

After Jerome's lecture from his dad, Laura turned to the rest of the room and paused a second to see whose loved one would come in next.

Because these stories aren't my own, I won't share any details, but the next two hours were profound. The majority of people who received readings had lost children. This was the first time I had ever seen the intense pain, intimacy, and rawness, as well as some healing, of this level of grief and unfairness up close. I don't think a day goes by that I don't have a memory of one of those readings that involves someone who lost a child.

Despite being deeply moved, the possibility of life after death was still very much at stake, and I paid careful attention to how evidential each point Laura made was, while doing my best to keep my head clear and neutrally evaluate everything she said. The possibility and my evaluation of an afterlife always took over beyond the emotion of the moment. I was still so shocked that the evidence kept building up and adding to the realistic possibility it was true.

I carefully evaluated everything Laura said. Was this someone she knew? Only one time during this day did she know who she read, and that ended up being a younger person who had not lost a child.

Was the information general—a boy who liked sports, or a girl who liked dolls? A small percent of it was, but not most of it.

Was it Google-able? Highly unlikely. A lot of the things she got were personal and specific. The odds of her having Googled them in advance and gotten that level of personal memories seemed almost impossible.

Did she ask leading questions, such as "You don't have a father on The Other Side do you?" Whether the person says yes or no, the medium then will answer either I thought so or I didn't think so. No, she didn't. Not once.

She did get some information wrong and, occasionally, offered information that was not recognized by anyone in the room, but those instances were only a small percent.

When the readings ended, there was a reverent silence in the room. Some people were in tears. Jerome was sitting there looking a bit stunned and a bit sheepish from his "stern talking to." I was still processing and digesting all of it. One minute I was in tears along with the parents, the next I was laughing along with them at a favorite memory of their child. We all *got* one another. We understood grief. We all understood a before and an after. There was an openness among all of us after experiencing such an emotional day together.

And then, of course, the overpowering fascination, relief, and hope. More evidence pointing in the direction of an afterlife. The fascination of a glimpse into another dimension as a normal person seemingly did the impossible.

I talked to a few people and we shared what has helped us cope.

I was careful not to tell any of them what had happened to me or reveal any identifying information about my family. Instead, I gave them names of the books and scientists that had helped me and commiserated about how brutally hard everything was now.

People lined up to talk to Laura, who I'm sure was exhausted both physically and mentally. But she still treated everyone she spoke with as if they were old friends. Her warmth seemed genuine. I continued my super-awkward lurking. She probably felt sorry for me and thought I had terrible social skills. I couldn't worry about that, though, because I needed to know how in the hell this totally normal-seeming woman very possibly communicated for two hours with dead people.

The room cleared out... aside from me and Jerome.

> **Me:** Hi! I don't know if you remember me, but I was at your last event. I know you worked really hard all day and wanna go home and all, but can I ask you a few questions?
> **Laura:** Hi, Liz. I do remember you! I am so glad you came back.

While I was surprised she remembered my name, her remembering me was clearly the least surprising thing of the day. She was certainly chipping away at my prejudices of what mediums were like.

> **Me:** How did you do that just now?! You know, it looked like you were really talking to dead people. I think you maybe even actually were! But HOW? I'm so jealous.
> **Laura:** You don't need a medium to talk to your people. You can do it yourself.
> **Me:** How? How would I know I was not just making it up? I wanna get "super evidential, scientists wanna study them" level signs. I would feel so much better. I tried asking for evidential signs like I learned about in your class and I haven't gotten them.

I was crying now. I was trying not to, but I couldn't hold back. Fuck. I just missed my dad so much.

Me: It also makes no sense because this person would always give me pretty much whatever I wanted. Oh wow. I'm sorry. That sounded so spoiled and obnoxious.

And shit. I realized right after I said it that it was too much information because that would eliminate a peer. So, my loss was now obviously an older generation. And very probably a parent. Although maybe a grandparent? FUUUUCKKKKK!

Laura: No, it doesn't. But maybe you need to give it a little more time. Write your signs down again.
Me: Okay. I will try. I... I can't say anything further, but like everyone else here, I had a terrible and traumatizing year. But I'm also confused about that card thing that happened. That was weird.
Laura: See? You have abilities.
Me: Uh, I'm still not sure I even believe that you have abilities. Oh my god! That was so rude! I'm sorry. I mean I am not sure I believe that anyone has abilities, that abilities are even possible. But... I dunno anymore. And that card thing—what was it? Did I remote view them or read your mind or was it predicting the future?

These possibilities were all new to me, and it was a bit thrilling trying to figure this out.

Laura: I'm not sure which it was, but it was definitely something. Did you contact Forever Family yet?

I had learned more about them. The Forever Family Foundation was started by a couple, Phran and Bob Ginsberg, after they lost their 15-year-old daughter Bailey in 2002. They started this organization in her honor to help grieving people by gathering evidence that showed consciousness can continue after death. They verified mediums through a series of specially designed tests and regularly collaborated with Windbridge. Their board included Dr. Diane

Hennacy-Powell, Loyd Auerbach and Dr. Jim Tucker. Like me, Phran and Bob were secular Jews, not very religious.

> **Me:** I went to contact them but saw you have to be a member for six months before volunteering. I think I will try the ASPR (American Society for Psychical Research) instead.
> **Laura:** You don't have to wait six-months. Just email them. Look up the email for Phran, and use my name.
> **Me:** Seriously? Thanks! I remember reading about them in your book, and the test they gave you sounded so interesting! I just… if I could just get enough evidence that all this is true…

I felt myself start crying again.

> **Laura:** I promise. It is. Your loved ones are around you.

I knew she believed that, but I needed more than belief.

On the walk to the train, Jerome called his sister: "Yeah. It was amazing. I think she is the real thing… I wasn't expecting her to be this good at all… Yeah, Dad barreled ahead of everyone and came in first… Yes… and he yelled at me that I needed to change my shoes."

He hung up.

> **Me:** She couldn't have Googled that, could she?
> **Jerome:** I never gave my name. Remember?

12

Five-Dollar Readings And Three-Hundred Dollar Candles

The ad said ten dollars.

I called the number and made sure to use my typical medium protocol. I blocked my number and withheld my full name. I knew this was total bullshit. But I had to see for myself.

What was the difference between a "phony" psychic and a supposedly genuine scientifically-backed medium? I had to find out.

Loyd Auerbach had warned us to never visit a storefront psychic, the performers and frauds who hang five/ten dollar palm reading signs and tons of crystal balls in the window. But that was exactly what I was stepping into.

If they did the things that Loyd had warned us about—hawking expensive crystals or overpriced candles to clear our bad energy—I wanted to see how this could actually work? How were they able to get people to jump from paying ten dollars to hundreds of dollars? Even thousands of dollars? Did they start off the scam with an accurate reading? Could they even be accurate?

The idea of an "upsell" wasn't what really mattered though. What really scared me was that they would be no different than the "real" mediums I had been researching. What would that mean for my investigation going forward?

She had no website, which was both different from every other medium I had been to and something Loyd had also warned us about: mediums should always have websites.

The differences between her and the "genuines" continued. Not only did I have to call instead of email or book through a website, she answered the phone immediately. And, when she booked my session herself, she didn't ask what kind of reading I wanted—Psychic or Medium. It felt very amateurish.

Also, unlike the other mediums I contacted, she was VERY available. She asked if I wanted to come in right away. I made an appointment for the following day. During our call, she was exceptionally warm, referring to me, again and again, as "sweety." She lacked the professionalism of the other mediums I had already met, but she didn't ask me for any information that would have tipped her off about me or my identity.

When I showed up to the appointment, I was taken aback. She was young and cute. I had been expecting an older woman with lots of new-agey adornments. Dressed in a semi-stylish green skirt and black tank top, she looked like she could have easily been in my social circle. She was friendly and seemed likable.

She held the session in the living room, which was decorated in the same Ikea-esque style I've come to expect of the apartments of my generation. Instead of crystal balls and beaded curtains, I stepped into a room of a typical post-college girl trying to look like she has a real home.

We sat down. Me on her couch and she on a chair facing me, the coffee table between us. She pulled out a laminated page with the prices and pointed to the psychic reading on the list.

Store Front: So, a full psychic reading is sixty dollars.

I glanced over the cream-colored paper with the crystal sketch at the top, with no price option below $45.

More differences. No welcoming. No asking if I had had a reading before. No explanation of what her abilities meant, such as whether she got her information clairaudiently (through hearing), clairvoyantly (through seeing), clairsentiently (through sensing), the way the "genuine" mediums usually love to educate.

Right to the money.

Me: What about the ten-dollar reading advertised outside.
Store Front: That's just for a palm reading.

I glanced along the page but saw no option for medium reading. Did she even know that there was a difference between a psychic and a medium reading?

Me: And what about a medium reading? If I want to talk to someone who had died?
Store Front: That's one hundred and twenty-five dollars.

To remain scientific, I knew I should get the same type of reading I had done previously with the other mediums, but I wasn't willing to drop $125.

Me: I only have forty dollars in my wallet.
Store Front: That's okay. There's an ATM downstairs.

I told her to start with the ten-dollar palm reading. She looked at my hand and began.

Store Front: You have a long lifeline and a good career, but you work very hard.

I live in New York City; everyone works hard here. She continued.

Store Front: You've had relationships that have hurt you. You are creative and you are trusting and have been hurt by friends.

Anyone else able to identify with any of the above?
I tried to look wide-eyed and amazed at how accurate she was.

She actually WAS accurate, but the issue was she would be accurate for the majority of humans. She was doing what I had learned is called a Barnum Reading. That is when someone describes general events, situations or personality traits that apply to most people.

A psychology professor had once conducted an experiment where he printed out an "astrological reading." He gave one to each student based on their star sign. Each student was amazed at how relatable it was. He then told the students to pass their reading to the student on the left. Each student had gotten the exact same reading.[1]

That was it for her ten-dollar reading.

I was too curious to let it go that quickly.

I firmly told her 40 dollars was my limit and asked what she offered in that price range. She agreed to do a "double palm reading." She had me put both hands palms up, pulled out three tarot cards, and began to read, expanding on what had happened in the pre-reading.

She had said only a few minutes before that I trusted everyone, but now she said that I didn't trust anyone or feel that I had anyone to talk to.

I agreed with what she said to keep her going. She told me the card said I would be married by 28 and done with money problems by 30. I didn't react.

She told me that I was good with money and good at saving money and, a little later, she told me that I needed to be better at saving because, in her words, "Money comes in one hand and out the other."

By offering contradictions, people who do believe and want to believe will remember the accurate parts and forget the inaccurate ones. She had her bases covered whether I was good with money or not.

She was also probably watching to see how I reacted to both to see if I responded to one more than the other so she could go further with one.

I didn't react to either.

She then asked what I did for work. I told her I do social media and branding for my own company. I did not tell her any details about my startup or that it was currently on hold.

Anyone knows that this would be a stressful job and most young women in NYC are incredibly driven and stressed out, so she took a very reasonable guess and mentioned how stressed I was and that a lot wasn't going right for me. That was age-appropriate, NYC-appropriate, and anyone-seeing-a-psychic-appropriate. People who sit with psychics generally go to hear about love, money and health. A "good" fake reading will touch on all of those.

She continued that I don't sleep well and that I was up late most nights. This was true, but it is also true for most people in my age group in NYC. I also probably looked exhausted because I was still in deep grief.

She then decided to go for it. In the middle of her generalities, she asked how long I have had that pressure in my chest when I wake up in the middle of the night.

I actually never have had that feeling, but I knew it was a pretty typical symptom of stress. I decided to lie and make her feel more motivated to go further—I made my face look taken aback.

Me: Um, for a few months.

She then went in for another score with the cards. Pointing to one, she asked who Matthew was.

I replied, honestly.

Me: My cousin.

The psychic asked what was going on with him. I said I would have to check. I don't know this cousin very well, but I did not tell her that.

She asked about the man in the uniform, obviously trying for another hit, a "reveal" that related to a large majority of people. I was honest and said I had no idea.

She covered her bases, though, and said that it must be someone I hadn't met yet.

She went back to how lonely I was and that I felt I did not have real friends to talk to, that all my friends were petty, caught up in gossip, and talking behind one another's backs. Again, a cliche of groups of girlfriends that is sadly often accurate. Although luckily it was not, I pretended that it was.

She then proceeded to tell me that I had a guy friend where it was kind of more, but that it would never happen because we just don't have that energy. I assume everyone has a friend of the gender that they are attracted to who, if circumstances were different, they maybe would be with, or if there was more of a spark.

She said I had had a failed relationship. Anyone else ever have one of those? That I had tried hard to make it work and the person had hurt me badly—twice. I assume I am not the only one who has been hurt more than once by the same ex.

I agreed with her. She said I still think about him. Yes. She asked me how old I was. I told her 28.

Store Front: The card said you were supposed to be married by this age.
Me: Oh great! So I will meet someone now?

She drew a horrified look on her face, as if I was in grave danger.

Store Front: No! You are supposed to be married already!

After planting the insecurity into me that I should have been married by now, which would have been cruel had I taken this seriously, she returned to talking about this ex. It *was* fascinating watching her engage in trickery. I knew I was witnessing a con artist in action. Even if she had bad intentions, human psychology is so intriguing!

I had been very deeply hurt by him and he had terrible energy. I had tried to make it work, but it didn't, which essentially describes every single breakup anyone has ever had.

The reason I was not married yet, she told me, was because it didn't work out with him. So she had managed to figure out that I am not married because I didn't stay with my ex and well... marry him?

She looked back at the cards, took some time to contemplate and let her message sink in before going back to it.

Her face contorted into a deep anxiety.

Store Front: Wow! You seriously should have been married by now.

She shook her head again.

Store Front: Things aren't working right for you because your energy is all off of alignment since you were so hurt by your ex. You have also trusted others who have let you down. Wow! Your whole energetic alignment is just so off.

She asked if I had a crystal. Here we go!

Me: Yes. I do.

I really do. Two actually.

Store Front: What size and kind?

I showed her the size with my hands and told her it was a clear crystal quartz. She then proceeded to ask if I had ever had a psychic reading.

I had had eight at this point.

Me: Ummm... well, I went to a group reading one time.

Her face contorted again, this time in disbelief, then disgust.

Store Front: You have! Really? Where was it?
Me: Umm, just this little shop. She seemed good?

I kept my voice insecure. I also tried really hard to hold back my amusement.

Store Front: What did she tell you?
Me: It was a big group. Not much.
Store Front: And she never told you how desperately you need an energy clearing?

It is rare in life someone follows a cliche and action you were warned about to such an exact degree. This played out almost exactly how Loyd Auerbach warned it would.

Me: Oh, umm, no. She never told me that.

She shook her head again.

Store Front: I can't believe that. You have all this awful energy mainly left over from how deeply hurt you were by your last breakup. It's stopping you from getting married, getting the money you want and having the life you want.

I did my best to look super worried.

Store Front: You seriously need candles and an energy clearing kit.

She happened to have one on her. For three hundred dollars.

Me: Ummm. Okay. I just need to wait for a few more work things to come through over the next month then I should be able to get one.
Store Front: Okay. You shouldn't delay too long. Please let me know as soon as you can.

To make sure I understood the direness of the situation, she filled her voice with deep concern. Then she ended the session.

For the sake of scientific consistency and my own curiosity, I asked her what I had recently been asking at every reading. I wanted to see how she would handle this differently.

Me: When you do a mediumship reading how do you know you are reading an actual passed person and not just my psychic energy thinking about them?

I stayed away from the lingo I usually used, like discarnate, so I did not reveal my sciencey-ness. So far, each medium had given me an intelligent answer. They either shared a personal experience with a client, a scientific study they participated in, or a fairly well-thought out response about why and how a medium reading, where they connect with someone who has died, is different for them than a psychic reading, where they read the mind and energy of a living person.

For many, it felt like just getting information during psychic readings, but actually interacting with a person during mediumship readings. Also mediumship was much more exhausting. A few told me about times they had gotten information about the sitter's deceased loved one that the sitter did not know, but found out later was true. So, they could not have been reading the sitter's mind.

She blew it off with a quick one line, obviously lacking any curiosity in her supposed "gift."

Store Front: It feels different. Different energy.

I was still chuckling as I headed out. The bell rang, and her next client came in, a sweet looking girl with a sad face. She had an anxious, maybe desperate, look. We smiled politely.

I no longer felt amused. I just felt sick and angry.

I headed home, unable to stop thinking about that girl who came in after me. What was she going through? How much money did this Pseudo- Psychic get from her? How far would she go? Had anyone come to her after the loss of their dad desperate for any hope that they were still around? After the loss of their child?

At least I did get to witness the huge difference between her and the other mediums, not only in terms of ethics but also how her general information clearly contrasted with the much more specific information most of the real ones offered.

But how many people never get to learn about this whole other level of scientific psychic-mediums because most think of fake storefront psychics like her? How many people never even get a chance of finding any valid hope their loved ones are still around.

13

Manifesting and Channeling

Laura Lynne Jackson was holding another workshop. This was a workshop on manifesting. While that wasn't actually about evidence of whether or not our consciousness survives bodily death, I was fascinated to see what any person who claimed to communicate with dead people, then backed up that claim with strong evidence, had to say about anything.

I also had a few important questions I needed to ask.

I took my seat. This time, I didn't look for recording devices or potential spies gathering information. I supposed I shouldn't have ruled it out, but I no longer suspected Laura was gathering information in any deceptive way.

She began the class by explaining how manifesting had worked in her own life. She clarified that manifesting did not mean that we didn't still work for our goals; it was more about what we put out and asked for, while trying to make them a reality. Along with doing the hard work we normally would, practices like making vision boards of our aspirations and focusing on intentions, writing letters to The Universe about what we wanted and asking our spirit guides would all increase the odds of achieving our goals.

I listened with an open mind. After everything I had seen, I was pretty open to the fact that I might not have any idea of how the universe actually worked after all.

As Laura gave examples of people who followed these exercises and manifested their dreams into reality, I was suddenly hit with a horrifying thought: What if I had known about these exercises when my dad was sick?

Could I have helped him recover? Would everything have turned out differently?

I pictured my mother rolling her eyes and saying, "That is exactly why religion is so toxic to people. It makes them think that if they prayed harder or loved God more, they could change things they couldn't. It's magical thinking and it's cruel."

As we cut out images of things we wanted to manifest in our lives from glossy lifestyle magazines, my hands started shaking. I felt as if I was going to throw-up over the idea that something so simple might have saved my dad. Yes, I knew that wasn't super logical and was very "magical thinking," but still… I needed to know.

I put down my scissors and ran up to Laura.

Me: Hi. I'm back.
Laura: Hi! So good to see you! Thanks for coming again.

I just stood there, staring at her.

FUCK! How could I ask this without giving away too much information? If I clarified that I had time to manifest about my loved one recovering, it would be clear that they died from an illness, not an accident. Actually, it would be clear they had not died from a heart attack, or any sudden ailment.

Also, if I was standing in front of Laura asking this question, I would naturally picture my dad and what happened. And maybe Laura would unintentionally read my mind. Then if she "brought in" my dad during my reading with her, how could I confirm she was really connecting with him? I couldn't risk the possibility that during our reading she wouldn't recognize my voice, consciously or not, and then remember this in-person conversation, also possibly unconsciously, and fill in the gaps. I tried to let my mind go blank.

WTF Just Happened?!

Me: Ummm… I don't know how to ask this…

Fuck! Seriously. How do I do this?

Instead of my mind going blank, it seemed to kick into overdrive. I heard the medium's voice in my head from my most recent reading. "Back and forth. He is telling me he went back and forth. Hospital. Rehab. Over about three months."

I froze. Don't let Laura read my mind now. I tried again.

Me: Okay, so there was this thing that happened, and I DID NOT want it to happen. Or depending on how you look at it, I wanted something to happen that didn't. But I didn't know about these techniques and, if I did, would it have turned out differently?

Tears filled my eyes and my throat tightened.

Laura: Was it related to work or something?
Me: No. Umm… it was so much more important than work. This was the most I ever wanted anything in my entire life. I guess I could say I wanted a different outcome.
Laura: Could it be fixed now?
Me: No. It's not fixable. I honestly do not know how to explain this and make sure I keep things evidential.
Laura: Can you tell me just a little more?
Me:…
Laura: ???

There was probably a social etiquette to how long I could expect her to stand there while I tried to frame my question. Especially in a class of other people who also wanted her time.

FUCK IT. I needed to know.

Me: I don't know how else to ask this, so I'll just have to say it. I wanted someone not to die. I wanted them to recover from either an accident or illness. And if I had known all these techniques…
Laura: Ah, okay. No, this couldn't have helped.

MANIFESTING AND CHANNELING

She explained that during our current lifetime, we have a certain number of "doors," or "chances" to pass away. These are potential points of exit. The number of potential exit points varies from person to person. And at some of these points there is no choice. It is when we have planned to cross over. These doors were decided by the "soul" in the life-planning process well before we were born. None of us can control, or have a say in, when another person dies.

I was openly crying. Again.

And, once again, Laura was incredibly patient with me.

> **Me:** So the rest of us are just stuck here to deal with it?! That ABSOLUTELY sucks. I am reading everything Julie Beischel and Ed Kelly and Sam Parnia and Stanley Krippner and Stephen Braude…

I spewed off the names of every doctor, Ph.D. and afterlife researcher's work I had been diving into.

> **Me:** But it still doesn't make any sense. It seems so… outlandish. Consciousness able to live outside of a brain! How? Where? By what mechanism? Or chemical composition?
> **Laura:** Have you gotten any signs?
> **Me:** NO! At least not anything completely undeniable like fifty feathers exploding over my head.

I had decided that none of my supposed signs were strong enough to be evidential. I needed more. Something bigger and more "not possible by normal means." And now I was worried that I shouldn't have told Laura that I had asked the universe to explode feathers over my head.

> **Laura:** I think maybe you're focusing too much on the science of it and not on the personal experience of it. Being so focused on the science won't answer as much as you would like.
> **Me:** THAT MAKES NO SENSE!

How would anyone study the existence of an afterlife in any valid way aside from science? And what did she even mean, "The personal experience of it"?

And that "points of exit" theory just brought on more questions: if someone has a soul agreement to die at, say, eighty, can they live self-destructively and drive drunk, smoke… What about things such as chemo? If you get cancer and are meant to die, can chemo change that? Modern medicine certainly seems to be pushing back a lot of death rates. Why does medicine work on some people and not others? And how did Laura know this? I know she said she gets all information from her guides, consciousnesses on The Other Side who help her. Did all mediums get the same information about these exit points?

I did at least feel a bit of relief that Laura said manifesting would not have saved my dad. Not that I necessarily believed that it could, but my definition of reality had been so seriously challenged over the past six months that I was not going to rule anything out.

I went back to my table and kept cutting out pictures of successful and stylish women at the top of their career, holding a baby next to their super-hot husband, minimalist Brooklyn lofts and girls in bikinis with very flat stomachs.

I started talking with the woman pasting her fantasy life next to me, when she turned it to what brought her here.

Woman: My grandfather passed away and my mom just got diagnosed with late-stage cancer.
Me: I am so sorry! I had a loss too.
Woman: I'm so sorry too. Who did you lose?
Me: Uhhhhh… please don't take this the wrong way because I am not a cold person or anything, but I can't tell you. I don't really think this, but I cannot prove that you don't know Laura and won't tell her.
Woman: Oh. I don't know her at all. This class is the first time I met her and I'm not gonna go hang out with her and talk about you or anything.
Me: Right. But I can't PROVE that.

MANIFESTING AND CHANNELING

Woman: Can I just ask you if it was a parent?

She was about to lose her mom. We looked close in age. She probably needed a feeling of camaraderie, something to show her that life was not only unfair to her. That others got fucked and dragged into the dead parent club before they should.

I looked around, then leaned into her ear and whispered, "Yes."

When class ended, people lined up to thank Laura. I lingered around until they were done so I could ask my questions. I narrowed it down to the two most important ones.

Me: Why do you think you are reading an actual deceased person and not just psychically reading someone? Why do YOU personally think this? I know all the studies at this point and how you see one on the left and one on the right of the screen, but I would like your personal opinion.

Laura: Okay. I had one client who lost his dad. His dad would sing him this song when he was a little kid. He wanted me to tell him the lyrics during his reading. I couldn't get them. Then a few weeks later, in the middle of the night, I felt his dad come to me and he gave me the lyrics. I texted them to my client immediately. He called me in tears saying they were accurate.

I had to agree that was pretty evidential.

Me: And I have another question.

It had just hit me a week ago that this could even be a possibility. An actual solution to something I had been deeply dreading, the upcoming holidays. My dad was always a vibrant life-force who made everyone laugh during holiday dinners.

I remembered Passover Seder, his eyes twinkling.

Dad: The joke is actually on us Jews. Here we are thousands of years later and we still have to eat this crappy, flavorless matzah cracker instead of bread. The Egyptians are probably laughing their asses off at us.

How would I get through a holiday dinner without him?! It seemed unthinkable.

Me: Do you have plans for the holidays?
Laura: Uhhh... which ones?
Me: Any. All of them actually.
Laura: Ummm...
Me: Would you want to... I wondered if maybe you could join my family for our holiday dinners?

I knew my mom would KILL me if Laura just showed up for our holidays, unannounced. And I certainly had no intention of telling her in advance considering I knew she would not be okay with it.

Laura: Uh, I'm sorry but... umm... did you... did you just invite me to join you and your family for the holidays?
Me: Yes. Oh wait—no. Not like that. Wow. I've met you like three times, and not even in a personal setting, so that would be actually super creepy. What I meant to ask was, could you come. We would pay you, of course. But is there a way you could physically join us but bring in someone who passed away so it would be like they were literally there with us but just in a different body?
Laura: Ah okay. Got it! No. I can't do that and I actually have plans with my own family on the holidays.

She seemed visibly relieved that someone she didn't even know personally, who kept showing up at all her workshops, hadn't actually invited her to join their family dinner as a friend.

Me: Right! Of course!

Of course she actually had her own family who she would want to be with on the holidays. Nothing like the self-absorption of grief!

> **Me:** But it doesn't have to be you. Is there anyone who can do this?
> **Laura:** That's called channeling. I don't know anyone, but ask Phran.

BY THIS TIME, Phran and I were already emailing and talking fairly regularly. After Laura's previous workshop at the Radisson, I had reached out to Phran to volunteer for Forever Family Foundation, just as Laura had suggested. Phran and I hit it off right away. She was energetic, smart, very funny and so passionate about her work. She had a strong Jewish, New York accent, which made me feel right at home, and she was "no-bullshit" direct. I immediately sensed I could trust her judgment and assessment of this unfamiliar and weird world of afterlife research and mediums. She had that instinct that immediately saw right through any nonsense and deception.

I told Phran about how Jim Tucker was the first one I had discovered and how that had changed my life and opened me up to exploring further. Next, I told her about my insane experience in Laura's first class when I "mediumed."

> **Me:** I honestly have no idea how it happened. I just guessed all this information and it was right? Trust me when I say I have NO abilities.
> **Phran:** Let me tell you, when someone's loved ones want to talk to them, they find a way. Even the crappiest mediums will, once in a while, suddenly get some real information if someone's deceased loved ones have been trying to talk to them forever. But here is the thing, and this is one of the reasons why the work we do at Forever Family Foundation is so important: do you know how many people

would go to Laura's class and have that happen to them and the next thing you know they are hanging up a shingle, calling themselves a medium and making people pay? Then people get terrible readings, they are devastated and think this is all a bunch of bullshit.
Me: No, I do not plan on taking clients. I truly just made up a bunch of stuff. But it sounds like you believe this is all true now? That consciousness survives?
Phran: Yes. Definitely. I have had so many experiences. But I always thought that there was something. Bob was the hard one to convince. It took years.
Me: And he thinks it is all true now?
Phran: Yes. Definitely. Wait until you start going further with this. So much starts to happen. You will see.
Me: I really hope so. I umm—I guess I should tell you what happened. Right? Who I lost? But can you promise me that you will never tell any of the mediums?
Phran: Don't tell me. Otherwise, whenever you get an evidential reading, you will always wonder if I told them.
Me: Thanks. Can I ask you another question?
Phran: Sure.
Me: Are you ever happy again?
Phran: Yes, you will be. It takes a lot of time and it is always going to be different, but you do get happy again.

THE DAY after Laura's manifesting class, I asked Phran during our Skype session about channeling.

>**Phran:** Channeling?! Who does channeling?
>**Me:** I don't know. I asked Laura if she could join us for the holidays to bring in our family member and have it be like they were actually with us, but just in a different body.

I still hadn't told Phran who I had lost.

Me: Laura said she couldn't, but that that is called channeling and to ask you who does it.
Phran: Ahh okay. That won't work.
Me: WHY NOT!?
Phran: First of all, no medium who's genuine can promise who will come in. If they claim they can bring who you want, they are full of shit. That's up to The Other Side who comes in. And anyone who would take you up on this request is full of shit. Don't hire them. You would be disappointed, and it would be a waste of money. I also don't know any medium who can channel like that. I can't suggest anyone for this and cannot recommend you do this.
Me: But…
Phran: If someone could, we would use them too.
Me: So what do I do?! How do I get through the holidays?!
Phran: I know it's hard. I locked myself in my room and did nothing the first few years. I still don't celebrate. You have to find your own way. There's no magic solution or way around grief.

14

Energy = Mindblown Squared

According to Einstein, energy can be neither created nor destroyed. Because energy seemed to be what the mediums said they felt and used to communicate with our loved ones, I needed to learn more about energy.

I Googled as much as I could about places where I could learn about energy—but in a sciencey way. I discovered the IAC, the International Academy of Consciousness.

They were either going to be amazing or they were a cult.

I took the risk of an attempted recruiting and went to an introductory class. It was held in a cozy classroom. There were three rows of six chairs each and only five other students. Books on consciousness and energy lined the walls. I felt like I was in a professor's office.

The teacher was a man named Michael and so far he seemed more the intellectual cool young teacher than a guru-esque cult recruiter. During this introductory class, he asked us why we were interested in the IAC. I avoided getting too personal and said I had had a loss and that I was researching to see if there was any chance consciousness could continue with no brain.

> **Me:** I would never have thought it was possible, but I discovered the work of Jim Tucker and was so blown away that I kept researching further and I guess coming here is part of my research.
> **Michael:** You like Jim Tucker!?

He handed me a flier showing that Dr. Tucker was participating in the IAC's upcoming conference in Florida. Unfortunately I wasn't gonna be able to make it.

Michael explained that the philosophy of IAC was not about wanting anyone to believe a certain dogma. The Academy simply taught what was out there about the study and emerging research of energy in a more evidence-minded approach than spiritual. They also conducted their own studies and experiments, usually on their main campus in Portugal.

He shared some theories of energy and consciousness. We live many lives and while we are in the material form we vibrate at a low level, but we also can go out of our bodies to visit other states of higher vibration consciousness. However, we cannot go too far while we are living even during Out of Body Experiences, usually called OBEs.

This all sounded a little out there, but no more so than anything else I had been studying.

> **Michael:** Does anyone have any questions?
> **Me:** Many. If after we die, we go to these higher states of knowledge and happiness, why do we even come back here in the first place?
> **Michael:** To learn lessons.
> **Me:** But it seems like when we are on the higher planes we already KNOW them, so then wouldn't coming back here be like if I decided at this age to return to kindergarten?
> **Michael:** We can't understand exactly. But you learn lessons here that you cannot learn there. Think of it this way, we would get bored if we just sat in bliss and never challenged ourselves. We

grow through the challenges we face here in ways we can't over there.

I liked that he paused a bit before answering me and thought about what I was asking. He did not seem to have a package of definite and ready-to-go answers.

Me: But if we are supposed to be more advanced and then come back here to learn, how come some of us seem to be not advanced at all? It seems as if you are saying that the energy of consciousness comes to grow and is curious, too, but then where does that darker need to harm come from? Are there dark energies out there too? Because those dark energies seem to thrive here.
Michael: Are you asking me to explain where evil comes from?
Me: Yes.
Michael: I don't know.

THAT SATURDAY, I headed back to IAC for a BioEnergy workshop. The first half of the class was a lecture, the second was exercises. Michael introduced us to some compelling theories of energy.

Subtle energy, which is lighter than air, is the life force that is all around us. There were different kinds of these subtle energies, too, like consciental energy, which is attached to thoughts and consciousness. There was immanent energy, which radiates from plants, trees, air, nature and cells in general.

We all first feel the energy inside us, but we can grow our awareness to feel energy outside of ourselves as well. For example: If you walk into a room of people who had been saying bad things about you, you feel this energy immediately. If you walk into a room of a close group of friends you have not seen in a long time, you feel that energy too. Usually, however, feeling energies is much subtler than that.

ENERGY = MINDBLOWN SQUARED

These exercises Michael taught would help us sense those subtler kinds of energies we encounter every day.

For the first exercise, we learned to project our energy outwards.

Carefully, I focused on the energy inside me. It felt a little bit like being excited or nervous. I felt butterflies move around my stomach, but they were more peaceful. Unlike being nervous or excited, there were no thoughts attached to this butterfly feeling, and there was also a feeling of warmth with a bit of pulsing. This warmth and pulsing built up and, as I focused on sending it out of me, I felt it kind of vibrate around me, as if it was occurring outside of my body.

Next, we reversed the flow to pull the energy back inside of ourselves. I felt it return inside my body in a strong warmth that seemed to find its way through my pores.

Michael directed us to hold out our hands, our palms facing one another. He told us to picture heat and energy growing between them. In the space between my palms, I felt a build-up of what I can only describe as a thickening warmth. He told us to throw this warmth from our right hand to our left, then back again. As I "threw" it back and forth, it felt as if it emitted from one palm and pulsed into the other. We then played with it, building it up and thickening it further between our hands. It was palpable and kind of doughy. As I pushed and pulled my hands together, it grew thicker, heavier.

I definitely was not expecting energy to feel so… substantial? As if it was actually… something?! I don't know what. I didn't want to think too deeply about it, though. I wanted to focus on feeling and playing with the material-ness of this energy.

It felt much stronger than imagination but not as strong as actual physical matter. What was so interesting was I had no idea what it was supposed to feel like. When we all shared the experience, most of us were consistent, and Michael backed us up saying that was how energy felt.

The hardest exercise we did was called a VELO. The goal was to move energy up and down our bodies, both along our skin and inside us through our organs. We sat in chairs and began by building the energy on the top of our heads. I pictured it and felt a heat that pulsed faster and faster, which I was able to direct into my head. The pressure built up inside my skull, intensifying, getting stronger and stronger. As it built up, I continued to focus on pushing it down, and the sensation seemed to leave my head, trickling down my neck and into my chest, although it was not as intense as when it was in my head.

But it stopped when it reached my stomach. It went away. I couldn't feel the energy anymore, and I became restless, agitated and claustrophobic. I didn't think I could sit still another second. The more I tried to feel the energy in my stomach, the more annoying it was.

I then skipped over my stomach and, when I focused on my legs, I felt the energy return and relocate there. My legs were not nearly as easy as my head, neck, arms and chest, but they were not as difficult as my stomach.

We were then instructed to direct the energy back up, slowly then more quickly. We held it in certain spots and pushed it out and sucked it back in.

Energy genuinely felt like a physical substance! A warm, buzzing substance that could move and increase by focus or thought. It felt too real and matter-like to fall into what I thought of as spiritual.

When I told Michael how I couldn't move energy through my stomach, he explained that was because of an emotional holding or a blocked chakra. It could be from a single traumatic incident or a lifetime of holding negative emotions there. Energy and chakras were like a hose, he explained: when you block off one spot, energy swells in other areas. Before the class, I would have blown this off as a bunch of nonsense, except that was exactly what it physically felt like.

We then worked on sensing other people's energy.

Michael sat in front of the room, leading us into a deep relaxation. He told us to pay attention to when he was sending the energy out, when he was absorbing it and if he was moving it clockwise or counter-clockwise. I could feel a slight difference between when Michael was sending the energy out and when he was absorbing it. When he was sending it out, the room felt a bit warmer and filled with a very subtle movement. When he stopped, the room felt a little cooler and still. It was a lot like the difference between when someone enters a room versus the emptiness they leave when they step out. Also, when Michael absorbed the energy, I felt faint pulses leaving from around me.

I couldn't tell when he was sending it counterclockwise or clockwise.

I thought about how our ability to sense energies is more innate as kids. We then learn to turn that off to be part of society and get through challenging or painful things. I can remember having a lot of sensations in my stomach when I was little, so much so that I had to numb them so I wouldn't feel sick. Like when I was struggling in my judgmental and competitive elementary school, where rules and "appropriateness" trumped self-expression and growth. As a kid, my "gut" let me know when something was wrong, but logic—as explained by my parents, "this is one of the best schools in the city"—overruled my intuition. I came to believe that was how you make good decisions. I had a lot to unlearn.

I had some intuitive understanding that if I could get back that ability to feel energies, it would attract the right people, people I had never met and opportunities I could not currently imagine.

At the end of the class, I realized that energy, when focused on, felt just as physical as skin and bones, was strong enough to bend spoons, and possibly move feathers. And maybe it was even able to communicate and feel and think and hold our personality after we die?

I went back the following day for a workshop on Out of Body Experiences, OBEs. We were told to dress warmly and to bring a blanket, because our body temperature drops during OBEs.

The teacher was a woman called Tricia. She divided the class into three parts: a lecture, subtle energy exercises and, last but not least, an attempt to go out of body.

During the lecture, Tricia covered the basics. When we project our "astral body" out of our physical body, she explained, it can reach different levels, from hovering directly over our physical body to flying to other areas on our planet, all the way to other planets and universes.

Sounded like bullshit, but I listened with an open and curious mind.

Your astral body, which Tricia described as kind of like a soul, remains connected to the physical body with what is called a silver cord, or a thick link of energy. If that cord is cut, you die. In fact, that is what happens when you do die. The cord breaks. But no one else can just go cut your cord.

> **Me:** What would happen if you can't get back in your body?
> **Tricia:** The silver cord keeps you connected. It really can't happen. It's much harder to stay out of your body than you think. Going back is an instinct, kind of like breathing.
> **Me:** What if someone takes over my body while I'm out and has me go murder a bunch of people or steal a bunch of stuff.
> **Tricia:** That can't happen. No one else who isn't attached to the cord can take over your body while you're out.
> **Me:** But how do you know all of this?

She explained that they had collected information over the years from many people who have gone out of body. When people have been experts at it, they have often gathered a lot of wisdom as well as practical tactics.

Tricia gave us tips on the best way to go out of body: make sure we were comfortable, have it quiet—even meditative music and relaxing sound bowls can interfere with our ability to leave our bodies. However, Tricia did use a white noise machine since the consistent whir drowned out any noises and did not engage and interest our brain the way music does. To improve our ability to go out of body, she told us, we should work with our energy as much as possible—absorbing energy, projecting it out and frequently practicing the VELO.

She let us know some things that can interfere with having an OBE: Fear, which was normal at first. Superstition, such as thinking you have huge abilities. And, lastly, if you tended to get super mindblown by this kind of thing.

Well, Fuck!

> **Me:** Umm... excuse me? What if I DO always get too mind-blown by everything?
>
> **Tricia:** Just try to calm yourself down. The more you do it, the less you will get shocked. Also, the more knowledge from a scientific perspective that you can gain will help. That knowledge helps bring the excitement down.

I could handle that.

We took down mats from a shelf and slipped under our blankets. Tricia dimmed the lights. We closed our eyes. With Tricia guiding us, we concentrated on our breathing, breathing in for a set number of seconds, holding it, then breathing out the same set number of seconds. She presented us with a few scenarios to visualize. I chose two: lying in a tub in the middle of the ocean slowly filling with water as I rose out of it; and an air balloon lifting me off my feet.

During the breathing, I felt myself relax deeply and my body grew numb and heavy as if I couldn't move it, but I knew I could. I had a slight sensation of rising, but it was far from intense.

As I relaxed, I started to visualize my body a few feet underground in a box. That could have been creepy because it sounds like a coffin, but I didn't associate it with any word or emotion. I felt myself rise up a few inches and watched the sides of the box lower around me, as if I were on an elevator platform that was rising up.

I looked up into a blue sky and told myself not to get excited, to just go with it. But then it stopped. And I was just lying there, back on my mat with my eyes shut. I briefly felt as if both of my arms rose a little, but neither actually moved. The second I thought about my arms, the rising sensation disappeared. I then lost all track of time and my mind went blank as I continued breathing.

When we were woken up, I looked around and felt surprised by the room I was in. I was disoriented and expected it to be larger with a higher ceiling and more windows.

I was the only first-timer; the other students all had more intense experiences of floating around and leaving their bodies.

Me: If I leave my body, can I visit those who have passed away?

Other students all said yes. Tricia backed them up.

Me: But how would I KNOW I saw them and didn't imagine it?

No answer.

Me: I guess I can have someone write down a number or place an object somewhere at a certain time and I'll tell them what I saw.
Tricia: Sometimes that works, but it doesn't always work like that. Your senses might not function the same way.

She explained that it could be more like a dream, where the place you see resembles reality but can have some changes.

In a later class, Michael explained that sometimes people visited places out of body and time didn't apply the same way. For example, someone went to a place they knew, but saw a large swimming pool that wasn't actually there. They then learned that a swimming pool had been in that spot 20 years earlier. People had also seen the future. One person had seen a playground in an area that was currently a parking lot. Ten years later, a playground was there.

Me: So, how will I ever KNOW for a fact I am OBE.

No definite answer, but there was some evidence. Sometimes two or more people would go OBE together and have the same experiences. Sometimes a person who was OBE would go and visit a person not OBE and see things that were later verified, or the person not OBE would say they saw the OBE person. And others HAD tried what I suggested and sometimes it worked. But not always. I later learned of more verifying stories when I took a class at The Rhine on OBEs taught by Graham Nicholls. He was once

OBE above a local neighborhood church and saw that there was damage to one of the windows and that window was covered in tape. The next day he walked to this church and saw there was recent damage to that window that matched his OBE.

I also later learned about an experiment done by the psychologist and parapsychologist Charles Tart. His kids' nanny, referred to as Ms. Z, claimed to have OBEs. Tart put a series of numbers on a shelf way above the bed in his sleep lab studio. He was able to observe her through a window to assure there was no cheating. She got the numbers right. He then tested her brain during an OBE, and the results showed patterns that were neither sleeping nor awake—in fact, it was a pattern he had never seen before.[1]

I left with the name of a book to read, Robert Monroe's *Journeys Out of the Body*.[2] Monroe was a "normal" businessman who attempted to verify his OBEs. He did, by all sorts of crazy ways, including once pinching a friend during an OBE who later confirmed she felt it.

As with everything else in this world, there were some eye-popping stories that hinted time and space were not bound by the laws we thought they were. I felt as if I was part of some massive treasure hunt and I was being given small clues. I imagined my dad and I putting these pieces together, both amazed. I had already come across enough clues to keep me guessing, and I knew something was going on, but I still had no idea what was actually going on.

15

Hiding Evidence And Meeting Mediums

I was delving into my work with Forever Family Foundation, promoting the upcoming Afterlife Explorers and Mediumship Convention in Florida. I was also going to attend. It was the first thing I was truly excited about in a long time. I would get to meet Phran and Bob, a bunch of mediums and Loyd Auerbach!

The morning of the convention, I woke up excited. And then I felt excited that I felt excited, then guilty that I felt so excited and then excited again.

It was November 10th, 2016. The country I had lived in my entire life had overnight fallen into the hands of a narcissistic, incompetent, orange-man at best, and a mentally unstable, evil dictator, at worst. Everyone on the streets, subways, airports, and cafes of New York was checking in to ask strangers if they were okay or to vent mutual rage or terror. I guess there are a lot of "What The Fuck" moments that are absolutely worldview-transforming. Only two days ago, November 8th, 2016, was one of them.

After my plane took off, thoughts kept turning over in my head. How would I handle landing and not being able to call my dad? This was the first flight of my life where I wouldn't be able to check in with him as soon as I landed. They say getting through the firsts

HIDING EVIDENCE AND MEETING MEDIUMS

are the hardest, and I had no idea how this first experience would hit me.

The other issue was that I would be spending the weekend with ten mediums in social settings. I had met Laura, but I couldn't tell her anything about me. I had had a phone reading with Eliza Rey so I wouldn't have to hide anything from her, but I would be meeting the others for the first time.

How could I have conversations when I couldn't tell them anything? OMG! I was going to be so awkward!

When I got to the hotel, I checked my phone to see what horrific nonsense our new Orange Overlord was bringing to the world, and what witty comments my network had about it. Settling into my room, I unpacked, checked my phone again, showered. I was actually happy to be here and eager to meet everyone in person. It felt so good to be happy!

I ran down to a large carpeted conference room, where everyone was stuffing gift bags. I instantly recognized Phran Ginsberg from our many video calls. She gave me a big hug. I got just as warm an introduction from her husband and Forever Family Foundation cofounder, Bob.

I joined a few other volunteers at a table as Phran began instructing us on what we would need to do to set up. I felt productive and useful; I was actually doing something again. I was also a part of something, and around people where I wasn't the one everyone felt sorry for. All of us had been through something awful and traumatic—that's what brought people to the Forever Family Foundation in the first place.

Phran turned to me and Leigh, the foundation's Vice President and Volunteer Coordinator.

> **Phran:** So you three are on the goody bags, and then Janet Mayer will be down in a second to help too.

Janet Mayer?! FUCK.

I had read about her in Dr. Gary Schwartz's books and I had just booked a reading with her under a fake name. What if she read my mind today and then recognized my voice, maybe even unconsciously? I wouldn't know then if it was my dad or her being psychic. I also had to make sure to watch everything I said and not slip up and give her any information.

Janet came in, hugged Phran and Leigh, then introduced herself to me. She seemed warm and welcoming. Before venturing into the world of mediumship, she was a florist. I had to admit, with her unassuming friendly manner, straight brown hair, simple and elegant style of jeans and white button-down blouse, she looked more like a florist than a medium. Despite being a very successful medium with many scientists writing about and researching her, she was humble.

We all sat down around the rectangular wooden conference table, stuffing pens and dream journals into small plastic bags branded with the Forever Family Foundation name and logo while making small talk. I had never factored in the possibility I would actually be meeting the mediums in casual settings. I had imagined that maybe I would meet them formally in experiments if I ever got a chance to help Dr. Beischel or someone at the ASPR or University of Virginia. Or I would continue to engage with them in structured workshops like Laura Lynne's, but I never imagined a scenario where we would just be hanging out.

What the hell was the social etiquette in this situation? How could I balance being friendly while still protecting my upcoming reading? Also I had my reading booked under a false identity. Was that rude? Was that disrespectful?

After we finished putting together gift bags and other setting-up stuff, it was time for a volunteer and medium dinner where Loyd Auerbach was slated to perform a magic show. As I entered the room, Phran pulled me over.

Phran: Loyd, have you met our social media girl Liz?
Me: Professor Auerbach! Hi. Nice to meet you. I have taken a lot of your online classes.
Loyd: Call me Loyd.

Me: Ah, okay. Loyd. Well, your classes were pretty transformative for me. They also really helped me cope with some hard times.

I wanted to ask him the question he had never directly answered: what did he actually think? Do we survive death? But we were piling in to get seats and it didn't seem like the best timing.

I grabbed a seat and a medium I recognized sat next to me. She was named Renee Buck. I knew she lived in LA and I let her know that I spend part of the year out there. I told her I was signed up for one of her workshops. Renee looked to be in her late 40s, but I found out later she was actually in her late 50s. She had light brown hair, bangs and she wore a pretty gold necklace over a floral dress.

Me: So anyway, I need to ask you... umm... did you know this was going to happen? That, ugh, sorry I can't even say his name... But did you know he was going to win the election?

I went on the assumption that Renee (as well as most people who live in LA) hated him as much as me and everyone I know did.

Renee: He actually didn't win. He got millions of fewer votes. And no. I thought it would be Hillary. This is awful, right?
Me: But what's going to happen next?
Renee: I'll see if I can figure it out. I wonder if this is what we need to clear out all the bad. It's as if the negative is rising to the top and coming out in some awful, but temporary, explosion. Anyway, on a less depressing note, I know you do social media for Forever Family, but is that what you do for work too?
Me: Uh... ummm...

Shit! Would refusing to tell her be rude?

Me: Well, I don't want to be rude, but I want to get a reading with you so I would rather not say.
Renee: Ah, of course. Don't tell me anything. Keep it evidential.

That was less awkward than I thought it would be.

I got a glass of wine, relaxed and began to meet some of the other mediums. It was surprisingly easy to have great conversations while sharing nothing personal.

After we finished eating dinner, Loyd began his magic show.

He did some mind-reading mentalist tricks that were captivating and impossible to figure out. Next, he took a group of gold balls and a group of blue balls, put them into a bag and without looking, walked around the room asking one medium at a time to take one and hide it in his or her hand.

> **Loyd:** Don't tell me who has which color and I'll read your minds and tell you if you have gold balls or blue balls.

At that moment, a huge laugh erupted from the back.

That was my first introduction to Rebecca Anne LoCicero. While I hadn't met her yet, I had read her bio so I could promote her on social: she was described as "bold, brassy, honest, loud, outgoing, accurate, hysterically funny and full of joy!"

> **Loyd:** Okay, Rebecca, I'll start with you. Lemme figure out what color ball you have.
> **Rebecca:** I normally give blue balls, I don't get them.

Everyone else joined in on the joke, but no one as enthusiastically as Rebecca. Loyd guessed a blue ball and he was right. Despite her insistence that she normally only gave them, it turned out Rebecca really did have a blue ball, and Loyd continued around the room getting each color correct.

The mediums all seemed as if they would be a lot of fun to hang out with. And while all of them appeared to have strong and unique personalities, not one of them was weird in the way you would expect.

As Loyd continued with more mind-reading mentalist tricks that kept us all mesmerized and baffled, everyone reacted with the same wonder.

"Wait! HOW ARE YOU DOING THIS? That is so weird." Their voices echoed around me in a chorus, but the loudest one stood out—

Rebecca: WHAT THE FUCK, LOYD!

Welcome to my life hanging out with all of these mediums! Minus the normal explanation at the end of it and all. But seriously… stage magic? That is what it takes to freak out psychic mediums? Stage Magic?!

THE NEXT MORNING, I joined the other guests who had just arrived as we took our seats in the hotel banquet hall for a welcome orientation. I looked around at all of the other guests. Even though I had not yet met any of them, we were all in this horror of grief and loss together and I felt an immediate camaraderie.

Phran and Bob gave a brief description of what the weekend would be about. We were divided into four groups for the breakout sessions we had signed up for: you lost someone and wanted to heal (Group 1), you have varying levels of psychic abilities you wanted to explore further (Groups 2–4).

I was in Group 1.

After each group session, we would all come together for medium readings and meals.

The opening talk was led by one of the mediums, Joe Shiel. I had not met him yet and I was intrigued by him. He was a medium and a spirit artist. That meant that along with giving readings, he also drew people's loved ones in spirit. I had examined his site to make sure the loved ones he brought in were not all biological since he could have just drawn resemblances. They weren't.

His talk focused on the ethics of mediumship and he spoke deeply from the heart. He was in his 50s, I would guess, and had a round, warm-looking face, a stocky build, and a dignified, calming presence. His serious and powerful voice vibrated through the room.

> **Joe S.:** You have to understand that this is not some parlor game. It is about working with grieving people in their most desperate of times. There is nobody more fragile than the grieving, and that is not something to take lightly.

I noticed that I had felt especially good since getting here, but I also felt a little strange. I kept feeling this odd buzzing excitement. It felt a lot like being a little kid on my birthday except that buzzing excitement also occurred outside my body, touching my skin as well as my insides. That must have been what healing and starting to move again felt like as my cells and nerves woke up?

Later that day, I ran into Phran.

> **Phran:** It is great having you here! You are so happy and have great energy!
> **Me:** Really? Me? But I'm still so sad. I feel like all I do is lie in bed and cry.
> **Phran:** Of course. That's normal. But you're a happy and positive person at the core. I can tell.
> **Me:** Am I? And will I ever be really happy again?

I had asked her that same question a couple of months ago on our first call, but I still needed reassurance. To actually be a happy person again felt pretty unattainable.

> **Phran:** Yes, but you will be a little different. You will always feel some grief. Don't believe the people who say bullshit like it takes one year to get over. No, it doesn't. You don't get over it; it integrates, but you will still be able to be happy and enjoy things.

I went upstairs to my room feeling relieved and a little disoriented and slightly off-balance by this buzzing excitement, but in a good way. I had brought a few cute dresses, so for the first time in almost a year, I put one on, did my makeup and headed down to dinner.

As I grabbed a seat at a table with the other guests I felt an immediate openness and a level of comfort I'd been missing over the past year.

Even though almost everyone ends up experiencing a severe trauma, we go about our social interactions as if we assume nobody has. Questions such as: What do your parents do? Do you have kids? A husband? Any brothers or sisters? Questions that are normally asked with the expectation those people were all still living were not asked here with that tone of assumption. No one here was privileged.

Right after dinner, Joe Shiel and another medium named Joe Perreta were slated to give readings to the group. In his late 20s, Joe P. was one of the two youngest mediums in Forever Family.

Joe P. went first. He accurately brought in a woman's husband and another woman's father, giving clear facts about them. Then he gave Loyd a reading! What a perfect person to observe how they conducted themselves while getting read. I had learned that Loyd and Joe P. had never met. In fact, many of the mediums had not met Loyd, nor one another until this weekend.

Loyd did look happy. I could tell the reading meant something, but overall he kept his face as stoic as possible and only answered yes or no. Although most of the answers were yeses.

Joe P.: I am getting that she was highly spiritual and maybe a medium herself. You worked together. I am not getting a full name, but I keep hearing Anmar—does that make sense?
Loyd: Yes.

I asked Loyd later: Was that Annette Martin?

Annette Martin was a medium Loyd admired and he had worked with her for years until she passed. They had also been close friends, and he often brought her up in his classes.

Loyd: Yes, it was.

Was it easier for mediums who have passed away to communicate with living mediums than it was for non-mediums who had passed away? Were there mediums on The Other Side who helped our loved ones reach out to us? That is still a question I have. No one really knows exactly what it is like there. It does seem some people who are on The Other Side are "better" communicators, but this is a question that researchers hope to study further.

Joe S. went next. He was fascinating to watch as he sketched the person he was verbally "bringing in." They came to life on paper for the audience and, most importantly, for their loved ones. He also worked differently than the other mediums I had watched give group readings so far, in that he did not walk over to anyone and bring in that specific person's loved one. He instead requested that no one claim the person until he had given a sufficient amount of evidence. One woman recognized her grandfather; and then when he did the next sketch another woman recognized her brother.

Joe S. then moved to his final loved one of the night: "I have a woman here. I have mid-40s. She passed in her mid-40s. An illness, but very quick. I have three. The number three. Three kids?"

My Grandma passed in her mid-40s from a quick and unexpected illness that spread into an infection. She had had three kids, but one had died at two years old.

"And whose birthday is November twelfth. This person is here for someone whose birthday is November twelfth?"

That was my birthday! I still waited for a little more information since I wanted to make sure it WAS for me before giving away all that information in a room full of mediums I had not had a reading with.

Joe S.: Is this for anyone here? Anyone?

I realized I had to go claim it.

Me: Umm—I think this sounds like me.
Joe S.: What parts do you recognize.
Me: November twelfth. This woman died of a sudden illness in her mid-40's and she had three kids.
Joe S.: Okay, I will continue with you.

He went back to drawing while speaking to me. I tried to focus on remaining a good sitter. Poker-faced and direct.

> **Joe S.:** She says you love to read. You are studying now. You read all the time.
> **Me:** Yes.
> **Joe S.:** And I feel a distance. You never got to meet her?
> **Me:** Yes.
> **Joe S.:** Because there was a split in the family or because she died before you were born?
> **Me:** Died before I was born.
> **Joe S.:** Okay, here you go. Do you recognize this drawing?
> **Me:** Somewhat.

It was an interesting drawing. It did not look like the photos I had seen of my grandma, but it looked a LOT like what my mom will probably look like in about 15 years. Maybe that is exactly what my grandma would have looked like if she had gotten older?

I was a bit overwhelmed. In a good way. This was really seeming to be real!

The evidence kept piling on. The thought that the mediums were cheating seemed less and less plausible, and the possibility of the universe offering hope of an afterlife was getting more and more real.

16

Ordinary People, Extraordinary Claims

After dinner and the evidential readings with the Joes, some of the mediums and a few volunteers went to grab drinks at the hotel bar. And I got to join! I could not wait to observe all the mediums in a comfortable social setting. I was still taken aback by how likable and relatable every single one of them was so far. It seemed highly unlikely they were delusional or deceptive. As far as I could tell.

I joined them in the hotel lobby bar.

I first sat with Renee, Laura, and the Joes. Janet Mayer was sitting there too. Fuck! The more she talked, the nicer she seemed. Giving her a fake identity at this point felt like it would be a betrayal. Should I go tell her the reading was me under a different name?

But just because they seemed really nice was not a reason to ruin my sciencey approach. Extraordinary claims require extraordinary evidence and "likability" wasn't extraordinary.

However, more drinks kept coming and we talked about everything from hooking up and funny stories to politics. And no. None of them knew Trump was going to "win," and while I definitely cannot speak for all the Forever Family mediums, all of the ones I talked to so far were not happy about it. That was a relief!

When a waiter brought our drinks and referred to one of them as Ma'am, she was noticeably bothered. "I don't look like a ma'am yet do I?" this medium who I am sure would prefer not to be named here asked me.

> **Me:** No! But we ARE in Florida. I think it's a Southern politeness thing. Not a commentary on skin tone or anything.

She was somewhat reassured. We then got into a discussion about how much we all hated that word and how it made us worry if we looked old.

So they were just as concerned with earthly vanity as the rest of us.

They were also a vibrant and warm group of people. I got caught up enough in the conversation that I forgot that they were supposed to be weird and that I was hanging out with them just so I could hang out with my dad. I was now just hanging out with them for themselves.

Anyone who spends time in LA knows the "essential oils cure a broken leg" crowd, but the mediums did not seem to be that way. Hanging out with them felt more like hanging out with my "build the wealth you want by improving your investor pitch deck and know startups take time to grow" friends over the "use this mantra to bring financial abundance by releasing the negative energies and subconscious blocks to money that are holding you back" crowd. They so far seemed a lot more down to earth and credible than I was expecting.

However, in the middle of normal socializing, little reminders would pop-up that these ordinary people possessed not at all ordinary abilities.

When Joe P. casually mentioned a good friend of his whose mom had died, Renee then jumped in and asked, "Oh was it of cancer and did her name begin with an A? And she always wore this necklace with a green stone." She had never met this person who lost his mom. In fact this was Renee's first time meeting Joe P.. Joe P. responded with, "Oh yeah," as if this was a typical exchange. (In their lives, it probably was.) The conversation then returned to normal as if nothing remarkable had transpired.

Me: Wait! What the fuck?!

They looked at each other and shrugged.

Renee: We get information like that sometimes.

Often, when having a drink or two, they explained their "abilities" would open up and information would "pop in."

While I usually tend to be really social and love to meet new people, this felt heightened and more intense than normal socializing. I was enjoying myself the way I normally do, but so much was at stake in what their personalities were like. It was hard for me to stay present and engaged in regular conversation. The more normal and likable they were, the more I believed they were honest people. And the more they became real people to me, the more arrogant it seemed to think, "Okay they are nice people and I believe they BELIEVE in what they do, but… " A layer of existential despair and grief was eroding by hanging out with them.

But we continued drinking and I was getting more comfortable just losing myself in the conversation. Eventually, I wound up sitting with Laura Lynne, who was just as likable to hang out with in-person as she seemed in her class, and Joe Perreta. Just as I had been feeling about Janet, I started to feel bad about the fake email and last name I had given Laura. I downed my drink and I finally decided to confess the truth.

Me: Okay, this is awkward. I have to tell you something and please don't be mad.
Laura: What?
Me: This is from a really long time ago, before I had met any of you in person. So I had found your name first on The Windbridge site and I am on your waitlist under a fake identity.

Although I had asked her in one of her classes if that was okay to do with mediums, if it would hurt my reading, I had not said I had done it to her. Also, I am not sure if her "okay" on that had taken into account my actually hanging out socially with any of them.

Joe P. started laughing.

> **Me:** Okay, I don't think you guys will Google me. I used to think that but I honestly don't anymore.
> **Laura:** Oh that's fine. Keep it fake. It'll be fun to see if I figure out it's you.

Joe P. was still laughing.

> **Joe P.:** We really don't Google you guys. I promise.

It was pretty clear from this interaction that they got this from people all the time.

> **Joe P.:** And, don't worry, I get it. I am actually super skeptical myself. I like to investigate everything.

Did he? How so?

> **Me:** But you know you have abilities right?
> **Joe P.:** Yes, but I still question and like to get verification I'm accurate.

I alternated between observing and joining in. Joe and I were close in age and both lived in NYC. I could see us maybe becoming friends one day? Until now I had some fantasy of what it would mean to be friends with a medium—given that an afterlife turned out to be true. I suppose this wasn't the kindest mindset, nor was it very respectful of the mediums, but I didn't really care who they were. Or, if I even liked them. Any medium would do for a friendship. I imagined hanging out with them would be me and them and then them letting me know when my dad joined in the conversation, as if he was joining a friend and me for lunch. I could just hang casually with my dad out of the formality of booking a session. Also, maybe the medium would even let me conduct some crazy experiments on them so I could keep gaining my evidence.

However, this mindset began to feel less relevant, and not so respectful of them as real people the more they actually became real people to me. So far, I was enjoying hanging out with them the way I did anytime I met new people I liked who didn't possess crazy "giving possible evidence that death is not final" abilities.

The conversation then turned to the "fact" that all of our deceased relatives are always hanging around us. Because, of course. I now had had enough drinks to ask one of the things I had wanted to know forever.

Me: So, if our people are around us all the time, what about when we are having sex?

They all casually looked at one another to see what the others thought and shrugged.

Joe P.: Eh. Yeah, sure.

I assumed their complete nonchalance must have meant they misunderstood me.

Me: Let me clarify—I asked if my relatives who have passed have been watching whenever I have had sex from the first time until now.
Joe P.: Yeah. I think so. Anyway, anyone want another drink?
Me: WHOA!!! Hold on! This is not okay! This is not okay at all!
Laura: Sure, they are around, but they don't care. They don't have a body anymore so they don't really think that way.
Me: But I think that way! And I care.

Laura and Joe then turned to some totally normal conversation about shoes or something that seemed completely meaningless, given the direness of what I had brought up.

I sat alone remembering EVERY SINGLE hook-up I ever had from high school through college—COLLEGE!—until now, from drunken frat parties to serious adult relationships in a totally different context than ever before.

> **Me:** Wait, you guys! Stop talking about shoes and dresses and stupid stuff. This is a REALLY BIG deal.

Apparently not to them.

We continued to drink more and laugh and have fun. They talked about their kids and husbands and boyfriends and ex-boyfriends. And not one line about a healing crystal or anything. These were not delusional people.

These were not "woo-minded" people. As with regular science, the one making the claims must matter.

17

Good And Evil

As I fell asleep after my second night in Florida, I noticed how much weirder I felt. When I climbed into bed, I felt as if that energy I had experienced earlier was stronger now, bouncing all around me and inside me. Was this excitement? A combination of healing and excitement? I had never felt this way before.

I fell into a deep sleep. It was one of the rare nights in almost a year that

I finally got a good night's sleep. When I woke up at 7 a.m., that weird buzzing energetic movement felt more extreme. It was actually a wonderful feeling. It ran through my entire body and even a few inches above my skin.

I decided to lay in bed for 15 more minutes and experience it. Then the most horrifying vision popped into my mind. I saw a clear image of Trump with a purely hateful glare in his eyes. A dark and cruel energy shot through me. This is exactly how I imagined the energy of pure evil would feel like. I sat up. Was that what it felt like to be a medium, heightened beautiful energies followed by evil ones?

However horrifying that evil energy felt, I loved the experience of feeling energies so intensely. If that even was what was going on.

I got up, showered (I was doing that again) and ran down to breakfast, where Rebecca was about to hold a group reading.

Rebecca burst into the room Rebecca-style, her sleeveless dress showed off her many tattoos. As usual, there was a tense anticipation in the air of everyone longing to get read, hoping that even if they didn't get read, her readings of others were accurate enough to give realistic confidence that life continued after death.

> **Rebecca:** Hi! If you don't know me yet, let me tell you a little about the way I work. I am honest. I pride myself on getting evidential information. I curse. And I won't bullshit you. I also need from you not to say anything more than a yes or no. Let them tell me.

She ran up to the first person.

> **Rebecca:** HI! I'M WITH YOU!
> **Woman:** Yes.
> **Rebecca:** He is fucking hilarious. He is the kind of guy I could see sitting and having a beer with and laughing the whole time. You two had a great relationship.
> **Woman:** (in tears) YES.
> **Rebecca:** Whoa. Okay, now I feel someone else.

She began to describe another man whom the woman recognized.

> **Rebecca:** Was this your ex-husband?
> **Woman:** Yes.
> **Rebecca:** But you were divorced before he passed? And you have three, no two kids?
> **Woman:** Yes. Two.
> **Rebecca:** Ugh, if you don't mind my saying it, I get why you divorced him. He is a huge pain in the ass! And his mom, your ex-mother-in-law, she is crossed over? And, if you don't mind my saying, your dad is telling me she was kind of a bitch.

The woman started laughing.

Woman: Oh my god! Yes.

The whole room was laughing, amused by the release at the taboo of saying the truth instead of idealizing those who have passed. And the laughing! It was okay to laugh even if we were grieving.

Next, Rebecca moved over to another woman. Without the woman telling her, Rebecca knew the woman had lost a daughter and she gently brought her in. Along with being outrageous and not holding back, she was also sensitive and respectful. In the middle of the reading, Rebecca raised her tattooed arms, which she always does, bringing her personality to her reading. This time, the lights directly above her flashed off and on a few times. No other lights in the room were affected at all. The entire room gasped.

Rebecca jolted back.

Rebecca: Did you guys all see that! I felt that! That electricity shot through this mic.
Woman: Oh my god! My daughter plays with lights all the time at home to let me know she is still around. She flashes them and makes them flicker.
Rebecca: And your daughter wanted to make sure you knew she was here today.

REBECCA WAS RUNNING the workshop I was headed to next with the other "I lost someone" people.

Again, she came in with her full-force energy.

Rebecca: Hi, everyone!! I want to give you a heads-up that I am very loud. I feel bad for the group next door, but they should have put me on the other side of the hotel. I'm here to talk to you very honestly about grief. And I hope you were warned about me. I am direct and I curse. A LOT.

Just as she had done earlier, she held true to all of the above. She dove right into the heart of grief, letting us know what we were feeling was okay.

> **Rebecca:** And regret. How many of you feel regret? I should have taken that call? Or not said that? That is normal. And guilt. People shouldn't be told not to feel guilty because it is part of the grieving process and, even if I did tell you not to feel guilty, would you not?

She might be outrageous, but she was also wise and caring and her honesty was refreshing.

> **Rebecca:** Look, we need to have some reality here—loss is for life and you will feel guilty the rest of your life, but you can put it aside too and feel other things and then go back to it.

Thank you! If one more person told me that I would get over my grief after X amount of time I would kill them. Or, even worse, were the ones who subtly, or not so subtly, implied I should be over it by now. This seemed to resonate with the rest of the group too. Everyone in the room was in tears. It was cathartic—Rebecca's acceptance of our experiencing grief on our own terms and the fact we all got each other lifted a layer of loneliness—this universal experience of grief. We ALL wished we could have spent that one extra day, not had that one last fight.

> **Rebecca:** We lose someone and fall apart. We can't stay in pieces, so we reconstruct. How many of us took time off after our person died? And you eventually went back to work, but are you the same? Are your goals the same? This reconstruction is part of grief and it will take your entire life. We grieve forever, so take your time with reconstructing.

I went up to her at the end.

Me: Thank you! That seriously was so helpful.
Rebecca: I like to tell it like it is. No bullshit. And I have to tell you since you are doing the social media and getting into this world — the energies you start to feel, especially as a medium, it's better than sex. Seriously.
Me: Uh… thanks for telling me.

THAT NIGHT at dinner Laura Lynne was giving a group reading. She had told me she was going to try to read me. I was nervous. As Renee and I walked into the banquet room together, I turned to her:

Me: Promise me you will cover your ears if I get read?
Renee: I honestly won't even remember what she says. When I read, I am tapping into another part of my brain.
Me: Okay, but you could be tapping into unconscious memories. Please just promise me you will cover them.

Renee joined the medium table at the back of the room, and I went to join a table of the other guests. I looked back to see which of the mediums were attending and they were ALL there. Every single one. FUCK!

I debated if it would be rude to walk over and ask all of them, aside from Eliza, to cover their ears if I got a reading, but decided against it. Even though I did want to protect my readings with them, I did not want them to hate me, and I sensed that would definitely be breaching some etiquette.

When Laura entered, the whole room got silent and stared at her with desperation. My stomach jumped and my heart started to beat super-fast. The energy in the room multiplied in strength, and every person sat up a bit straighter. It hit me the level of pressure that must be for the mediums, having so much pain riding on their shoulders.

There was another level of intensity at this dinner, a large banquet room filled with about 100 people, because many people who were not here for the whole weekend came just to get a reading with Laura. While it seemed she gave excellent readings to everyone, she specialized in people who had lost a child. Almost everyone in the room was a grieving parent.

Laura went over to the first table she "felt pulled to" and began to tell this group of grieving parents who she felt come in (from The Other Side). She spoke to each parent for about 10 minutes, relaying information about their children. Her reading was once again astoundingly detailed and accurate.

She spoke to parents who had lost children and young adults in accidents, to childhood cancer, other diseases, and even to murder.

I watched their faces as each parent or set of parents got read. They looked first amazed and then would get a flash of happiness and hope while the panic relaxed as Laura got more and more accurate information. A heaviness and greyness would lift off of them and I could get a glimpse of what they must have looked like before the loss of their child.

She knew ages, jobs for those who were older and favorite subjects of those who were younger. She also gave many names, personality traits, quirks and physical characteristics.

Hearing the stories of families who have lost a child was becoming common, but no matter how many times I listened to a medium do it, I could never get used to it.

The graceful way Laura approached brutal deaths of young people was a powerful lesson in how to gently communicate with people. It was a highly delicate balance between giving the accurate details of how these kids (or young adults) had passed, without ripping open their wounds. All of the mediums this week managed this beautifully, but another level of care and delicacy was needed when it was a young person who had died and the message was for their parents.

That weird energy I had been feeling all day was growing stronger; it was warm and vibrating to a beat as if heat and waves were moving and swirling around my body and then up and down through me. The waves, which felt like they moved at different rates, were accompanied by powerful emotions that went up and down with the evidence and information she gave.

I tried to remember if I had ever felt anything like that before. Maybe it was a bit like a ride in an amusement park, but one of the gentler ones. More like a carousel instead of a rollercoaster. It also felt as if it was happening outside my body as well as inside. It was as if there was a field of nerves outside my skin identical to the ones running under it.

Laura then came by my table and I stared up in anticipation as my heart began pounding and my palms began to sweat. While I willed my dad to come through to talk to me, I tried to stay composed.

Seeing me staring at her longingly, she made eye contact with me and gently said, "I'm sorry my guides are now telling me the next few readings will be for people who lost children."

From what I was learning, it was not up to the mediums who they read. It was up to The Other Side and/or their "guides" on The Other Side. It must also be hard on the mediums to not be able to give everyone a reading when all of us there were hurting so much. I was disappointed. I felt the disappointment run through my body, inside and just outside of it.

These next two readings were especially heartbreaking. One woman had lost two young kids a year apart, and the other woman had recently lost a young daughter who was her only child. During her reading, she broke down as if she had been trying to hold herself together for a long time.

After about three hours, Laura announced she was done. The room sat still in awe-struck silence for a while after she left. I was overwhelmed with wonderment at the level and quantity of evidence, as well as the deep emotions of the parents. A few of the guests and I began to talk about what we had just witnessed.

Woman 1: Can you believe those lights?
Me: What lights?
Woman 2: The flashing lights.

Me: What flashing lights?

Woman 2: Oh, you missed it because your back was to the table. The lights over the table of mediums was flashing the entire time as soon as Laura started and stopped as soon as she was done.

To this day people still talk about the flashing lights at Laura Lynne's reading in Florida, and my back was turned. DAMMIT!

I was still reeling from the emotions and level of evidence as I walked to the elevator to go back to my room and freshen up before drinks. Laura happened to be getting in at the same time. I felt bad for her because she was understandably depleted and was probably scared I was going to corner her and bombard her with questions. I managed to contain myself.

After her mind-blowing accuracy, coupled with the power it had over everyone and the inexplicable physical sensations outside my body, normal explanations for what was going on were making less and less sense. After all, how could consciousness exist with no brain, but there was Laura once again defying logic and the laws of the universe.

THAT NEXT MORNING, I woke up and ran down to breakfast, still disoriented from the craziness of the past few days. I was also tired because there was a false smoke alarm that went off at about three am and we all had to evacuate our rooms.

I knew who to hunt down.

Rebecca: Social media girl! It was NOT my fault. Everyone is coming up to me like, "fucking Rebecca!" I promise my energy is not THAT strong.

I had learned she could flash off lights, cause electronics to get really buggy, and even once blew out the video equipment during her test for Forever Family Foundation while reading the sitters online. But apparently she could not affect fire alarms.

WTF Just Happened?!

After breakfast, we had another hour until the first activity of the day. I saw Loyd enjoying some peaceful downtime by himself.

Me: Hi!
Loyd: Hi. So, what did you think of Laura last night?
Me: I have no idea anymore. She… she was mind-blowing. I really wish someone like Michael Shermer or Stephen Hawking had been there. What would the skeptics and scientists have said?
Loyd: There are ones who still would not believe she was genuine and would say she used fraud.
Me: But… HOW could she have? After seeing her and everyone this weekend actually, I literally cannot figure out what even some hardcore skeptics like Penn and Teller could say.
Loyd: They would say that she hired a detective to do background checks on all of the people. Got their names and info from when they signed up for the dinner.
Me: But ALL of them? Like all fifteen? Those would have to be some pretty detailed and expensive background checks. I think that could cost more than she makes in a year. And the information—the personal info she got could not be found in a background check.
Loyd: She would have had to hire multiple detectives and had them research all these people's social media and hack their emails and talk to their friends.
Me: But wouldn't their friends tell them if some person was poking around and asking questions about them? And the mediums don't get the guest lists. How would she get their names without Phran and Bob knowing?
Loyd: They would say either Phran and Bob were in on it, she hacked their emails, or she somehow tricked them into giving it.
Me: But…

I had thought of all those things myself but none of them made sense.

Loyd: They will say anything in the world rather than acknowledge what happens outside of their sense of normal.
Me: But even then she could not have gotten those personal details.

As I said this, the profundity of what I was realizing washed over me. The pieces of all the research I had been doing started to come together and the full picture it formed took my breath away.

Me: Take away what I know of Laura's personality so far and the fact that she seems like she never WOULD deceive anyone, there just seems to be no way she COULD have done what she did by normal means.
Loyd: I would agree with you.

I stood there with one of the world's best parapsychologists and stage magicians who knew all the tricks of fraud and illusion that Laura could have used, and the further we delved into how she could have done it, the more unlikely and absurd our explanations became.

Me: What do you think? Personally, I mean. Was what Laura did psychic or super psi—you know, reading it from some stored bank of all that has ever happened—still out of the known laws of the universe but not actually our discarnate's consciousness. Or do you think there actually is an afterlife?
Loyd: Yes. I think there is. I have come to conclude we do survive death.

So he *did* conclude, after all that he had seen, that we survive!

I FELT a lightness and joy as I ran to make it to the first activity: Joe Perreta would be holding a workshop for the "people who have lost someone" group.

I ran in and grabbed a seat in the front row. As I had mentioned, Joe P. was one of the youngest mediums and he reminded me of someone who would be in my friend group. Every time I thought about it, I still felt taken aback by the mediums' "normalness." This talk was about dreams and how to use them to connect with our loved ones.

Joe P. sat comfortably on the desk in front of the class. He had a slight Long Island Italian accent, dark brown hair and large brown eyes.

He opened the workshop by talking about normal dreams versus visitation dreams. A visitation is special, because that is when our loved ones' disembodied consciousness actually communicates with us. We are not making them up like we do in a normal dream.

He shared that he had lost two of his grandparents a year apart.

Joe P.: Mediums still grieve. It is much harder for me to connect with my own loved ones than the loved ones of someone I don't know.

He had dreamt of them a lot, but he didn't remember the details or feel that affected by the dreams until one night when his dream was much more vivid. He woke up filled with emotions and he felt deeply connected to his grandmother.

Joe P.: I still get chills when I think about it. That feeling you get from a real visitation is like nothing I can explain. It's intense. It stays with you.

That WAS the huge difference I noticed in the "visitation?" I had had of my dad. Maybe that was really my dad?! I felt a jolt of excitement and a few chills of awe. If I really had communicated with my dad that one time, maybe that meant I could communicate with him again. I needed to learn more.

Joe P.: Does anyone have any questions about how to optimize this?

Joe P.: Yes, Liz?

Me: If you wanna ask a specific question that would prove the survival hypotheses, can you?

Joe P.: Yes, you can. Wait—that is so weird. I just felt someone touch me. That doesn't happen to me normally. That's a little scary.

Another Guest: It sounds like her people are telling you to get it answered.

Joe P.: Yes, I should! Are you talking about a validation? Show me this or that?

Me: I want to ask them to tell me something that I would never know, but someone living knows. Then I can check with the living person and know the dream could not have come from my own subconscious.

Joe P.: Ohhh. Okay. That is different than what I thought you meant. I thought you meant guidance. What you are doing when you say that is like saying, "I don't believe you are here."

Me: I don't!

Joe P.: Right. I am a skeptic too, so I get it. But I would ask for something not so specific. But you can try. If you only want one thing and then, if it doesn't happen, you could get caught up in that and miss other signs. They're still people and might not do exactly what you want.

Me: I hope that they're still people.

That also didn't make sense because why would my dad not do his best to prove he was still around? He was one of the ones who helped teach me there was no afterlife while here, so wouldn't he want to undo that if there was one?

I "spoke" to my dad in my head: "Dad! Come on. Get me this next level of evidence in a dream. I need to KNOW if there really is an afterlife."

After Joe's workshop, Laura Lynne's workshop began. I noticed I felt even weirder. That wavy energetic feeling was the strongest that it had been so far.

Laura guided us into a meditation and told us to communicate our signs with our loved ones. I found it relaxing, but that weird heat and energy kept buzzing and waving more intensely than ever. At the same time, a wave of sadness overtook me, which felt like a huge aching nausea minus the need to throw up. The fact that my dad was truly gone hit me. They call them grief waves and I hate them.

As I tried to ask for a sign, I kept picturing boats and that one time I saw the boat with his name, even though I kept trying to picture other things and get my mind off of boats.

Me: (to my dad) Can you just send me real evidence that there is actually an afterlife and that you are with me? How about having a pink and purple spotted 100 pound turtle walk in… now!

That didn't happen.

Me: Okay fine. So what could be indisputably evidential, but also still within the laws of the universe so you can do it. FUCKING HELL! CAN YOU JUST GIVE ME REAL INDISPUTABLE EVIDENCE?

Then it got weird.

The heat above my head intensified and seared from my crown into my body. In a shot, words came to me along with an incredible energy. I didn't actually HEAR the words in my ears; it was as if they poured into me out of nowhere, not through my normal physical senses. They were in my dad's voice.

JESUS CHRIST! HOW MUCH FUCKING EVIDENCE DO YOU NEED?

That was exactly the kind of thing he would say. All of the signs I had gotten so far came shooting into my mind.

I remembered how my brain had felt while receiving those signs, which I realized was a much milder version of what I felt now during this shot of energy. All these thoughts came at me at once within a split second, but were clear and distinct.

As if the same laws of time didn't apply?

I instantly felt transformed. My sadness and frustration were washed away, replaced by the amazement of the signs I had already gotten and the fact of how hilarious it was that my dad "said" that to me.

I couldn't stop laughing.

TO CAP OFF THE CONFERENCE, I attended a workshop with Eliza Rey, who led us in a journaling exercise. I had so many thoughts circling in my head and I still felt wavy and strange and really good. I was still processing and absorbing everything I had seen this weekend from the evidential readings, to my conversation with Loyd, to hanging out with the mediums when they were not "on." While I could not say I was one hundred percent convinced they were actually communicating with dead people, the idea of them as frauds or delusional no longer seemed to fit either.

The shock of that realization sent chills as well as a bubbly joy through me. When Eliza asked us to share aloud from our journals about what this weekend had meant to us, I shared one of my favorite quotes from William James based on his research on the medium Leonora Piper. Almost 100 years ago, he had ultimately come to the same conclusion I was starting to:

> *"I should be willing now to stake as much money on Mrs. Piper's honesty as on that of anyone I know, and I am quite satisfied to leave my reputation for wisdom or folly, so far as human nature is concerned, to stand or fall by this declaration."*

Without fraud or delusion what was left? Survival? Something else? I still had so much more to explore.

18

The #1 WTF To Top All WTF's That Ever Were

There was a final dinner. A handful of guests stayed on to attend. It was called "Strut Your Stuff" and it offered guests on the "I have abilities" track a chance to demonstrate any mediumship abilities they had developed over the weekend. Eight of the mediums sat at a panel table to coach. I took a seat with Phran and Bob.

Bob: Hey, Liz. Are you going to go up and demonstrate your abilities?

He was obviously joking.

Me: Ha! No Way!

I suddenly had a funny idea.

Me: You know what? I took a class with Loyd and we learned about cold readings.

I turned to Loyd, who was sitting one table over.

Me: Hey. I'm gonna try a cold reading!

Phran then explained cold reading to the guests, stressing that it has its place at parties and magic shows, where everyone knows it's entertainment, but it is really evil if done to trick people into thinking it was real. Especially those in grief.

I stood in front of the group and, playfully, and pretty poorly, imitated how mediums speak. When I started, I felt a little bit funny. Those same waves were moving more strongly through me—what I still assumed was the sensation of my brain neurons healing from a year of deep grief and hiding away from people.

Me: I am getting someone with a name with J or G sound on The Other Side, maybe John? A grandfather energy. And they are sending love. They definitely loved you and want you to know. Does this make sense to anyone? Does anyone have a grandfather on The Other Side whose name begins with a J sound? Who loved them?

The odds of that applying to more than one person in a crowd was very much in my favor. Four of the mediums raised their hands. I moved towards their table and looked at each medium, one at a time. The buzzing around me seemed to grow stronger, and the waves rolling around my body and through my head intensified and increased in frequency. It felt a little like I was bobbing up and down, like being on a boat. My head began pounding. It felt kind of like a fun, buzzy drunk, but I was still clear-headed, though I didn't know what to make of the pounding in my head. It felt good in an interesting way and I was having fun. I turned to Laura and started reciting facts about her life from her book, knowing she would get I wasn't really "reading" her with some magical powers or anything.

Suddenly, I froze. I realized in the middle of these weird spinning and pulsating waves that I did not want to do this. It felt very mean to use someone's real-life experiences for entertainment. I was scared I was going to say something hurtful. I stood there trying to see if I could come up with something I had read in Laura's book that was not about a loss or something that she might find sensitive.

My mind went blank. Even though I tried, I couldn't think of anything. At all.

I lost all sense of time. I could have been standing there for ten seconds or for two hours. I had no idea. You would think freezing in front of a crowd like that would feel awkward, but I didn't feel awkward at all. I just felt the wavy buzzy pulsating warmth get really, really strong. The increasingly hot waves felt as if they were pouring through the middle of my brain.

I then forgot about Laura's book altogether and I walked over to Joe Shiel, who had also recognized a few of the things I had said. I am not sure why I chose Joe. There was no thought process behind it. I just did.

I started to say anything that came to my mind. The information felt as if it poured into my mind from outside the top of my head in waves of tingling heat.

> **Me:** I had brought up your grandfather. You were especially close to him?
> **Joe S.:** Yes. He raised me.
> **Me:** And you have a brother who is still living?
> **Joe S.:** Yes.
> **Me:** And he is younger?
> **Joe S.:** Yes.
> **Me:** WAIT. WHAT???
> **Joe S.:** Yes. This is all accurate.
> **Me:** Umm… and water like an ocean or a lake was very important to your grandfather and to your childhood?
> **Joe S.:** Yes, I live on a lake.
> **Me:** And there was a boat that was very special to you guys and there is one day that stood out above all the others.
> **Joe S.:** Yes.
> **Me:** WAIT! IS THIS A JOKE?

Joe remained calm and sincere.

Joe S.: No. It's not a joke.
Me: BUT... HOW? HOW AM I DOING THIS?! Okay, I just I don't know why, but I thought of Christmas. Christmas meant something really important to you guys. But something different than typical Christmas.

Joe S.: Yes.

He was SO calm and steady.

I looked around at the rest of the mediums. They were all gazing at me in surprise and amusement—all of them, plus Phran, Bob and Loyd.

Me: ARE YOU SEEING THIS? WHAT THE FUCK!!! WHAT IN THE ACTUAL FUCK!!!
Rebecca: I LIKE her. She said fuck.

She was recording me on her phone.

Me: Rebecca! You can't show this to anyone!

I was so startled I couldn't stop laughing. Joe just kept looking at me while staying calm and steady. And then it got weirder.

The heat at the top of my head increased, poured down through my head and neck and consolidated in my chest, where it started pulsating intensely. The waves in my chest got hotter and pulsated faster, then burst and built up in my chest before bursting again. This kept repeating.

Me: Umm... I'm feeling this intense bursting heat in my chest? Did one of your grandparents die of a heart attack?
Joe S.: Yes. On Christmas day.
Me: SERIOUSLY!?!
Joe S.: Just keep going. You got this.

How was he staying so calm?!

Me: Do you promise this isn't a joke?!

But I knew it wasn't.

The pulsing in my chest stopped and moved to a single spot on my head. The pressure built up, then burst and built up and burst again. This burst kept repeating in the same spot in my head.

Me: And one of your parents died of a brain tumor?... No. No, wait, this is a burst, Was it an aneurysm?
Joe S.: Yes. But not my parent. Try someone else.
Me: ...
Joe S.: My grandmother.
Me: THIS IS SOOOO WEIRD!

Joe remained calm.
I did not.

Me: And were their initials C and E?
Joe S.: No, those are the initials of my kids.
Me: REALLY!? WHAT IN THE ACTUAL FUCK!

During this whole experience, I could hear the mediums trying to give me advice, but I could not hear what they were saying or who was saying what. Everyone felt very far away, except for Joe and Rebecca, who was seated next to him, declaring her like of me for my use of the word fuck.

Joe nodded, calmly and respectfully. He acknowledged each piece of information as accurate. Then I was done.

Joe thanked me to let me know it was time to move on. I had nothing more to say anyway, but I still felt that weird buzziness all around me.

Me: HOW DID I DO THAT?

THE #1 WTF TO TOP ALL WTF'S THAT EVER WERE

Before I went to take my seat Joe handed me a note. He had written down everything I said and how each thing applied to him. I took it with trembling hands, then rejoined everyone at my table. As the hot pulsing waves started to subside, I was still shaking and absolutely stunned.

After dinner, I went to the bar to hang out. My whole body still felt weird and wavy, although not as strongly as during my "stint as a medium." Laura Lynne joined me at the bar. I was trying to calm down and stop my hands from shaking so that I could sip my drink without spilling it everywhere.

Before Laura could grab a drink, I accosted her.

Me: WHAT THE FUCK! What was that?
Laura: Ha. I have been telling you this stuff is real. You just needed to have a personal experience.
Me: But… uh… but… science… how!?

I tried to reply like a coherent person. Unsuccessfully.

Laura: And you got that information on me too.
Me: Uh, yeah? I read your book.
Laura: You got some stuff that wasn't in my book.
Me: Wait, WHAT? Okay stop. Not funny.
Laura: I'm not joking. I promise.

Loyd came to join me.

Me: What in the fuck was that?

He was still laughing.

Loyd: You are a terrible cold reader. You do know that was NOT a cold reading.
Me: APPARENTLY!
Loyd: You got real information. There is no normal way you could have gotten all of that information.

He then said something I badly needed to hear. Not only for this night but for everything I had been experiencing so far and would end up experiencing in the future.

> **Loyd:** Don't ever rationalize this away or tell yourself this never really happened. Because it did.

19

I'm Not Okay. Okay?

He was middle-aged, slim, elegant and stylish. He looked more like an art director of a trendy ad agency than a medium. He spoke crisply and politely. He knew I had just come from a trip. He brought up how I had taken a car service to the airport and that I had flown on JetBlue.

This was now my 21st reading.

He started whispering dialogue, as if he were getting it from my dad.

Medium: He's glad you were laughing.

Then he turned to me.

Medium: He says you were having fun and laughing. That it was great to see you laughing again.
Me: Yes.
Medium: He was there with you. And then he gave me the feeling of a performance. You did some performance. It feels like it was not Broadway-style or anything, but it was something that feels really, really out there. And it feels like it is out there compared to

your dad's sensibilities. And it was really out there for you, too. Whether he liked it or not is not the issue, even if it was too out there for him; he wants you to know he was with you.

I started laughing.

Me: Wow. Yes.

I sat through the rest of the session with a medium who may have thought I had done some super risqué sex performance before I got to explain what I actually did at the end of the reading. When I told him what had happened, he laughed, then explained the sensations I had experienced.

Medium: Mental mediumship is controlled by the central nervous system and people in the spiritual realm will use your central nervous system to convey information to you. The spirit exists at a higher rate of vibration, and they impress on the nervous system their energetic vibration in order for the medium to be able to convey the information. The more often we do this work, the more neural pathways we create for the information to go through.

He was the first person to put it this way. I think that was the most sciencey, logical explanation I had heard at this point for something that had been far from explicable with science and logic. Again, this showed that our brains and neurons grow, develop, then filter based much more on our culture than on reality.

Which made me wonder, what percent of reality did we even perceive?

After I got back from Florida, everything felt different. That wavy buzzy energy lasted for about five days, although each day it got a little weaker. Once I had a personal insight into how a medium experiences giving readings and got a taste of how they felt physically, a lot of what I had been studying fell into place.

All of my readings ended up being different from this point on. They were more collaborative and I was able to identify with what they were experiencing. At least somewhat. I started to trust them more. When they shared that they were confused about something or that they did not know why they had a certain physical sensation or saw a certain image, it no longer sounded as if they were making that up. I worked with them to figure it out, but not so much that I gave too much away.

I still wanted my evidence!

At the same time, I was dealing with a selection of friends who did not get how I handled grief. They kept trying to engage me in the world in a way that would make them (not me) feel better. I didn't feel better. I felt like shit. One week after Florida, the texting picked up again from one of the most active Grief Conversion Therapists (GCT) in my life. Grief Conversion Therapists, as I had stared to call them, are the grief version of Gay Conversion Therapists. They want you to change to adjust to their definition of okay, regardless of whether that is what you need.

Some people were wonderful, and I do know people often have no idea what to do and genuinely want to help. In the depths of my grief, however, I needed my process respected, not corrected. I just wanted people to sit with me where I was at and tell me it was okay that I would feel awful for a while, not as if I needed to either cheer up or at the very least present as cheered up.

Some did this and some did not. Some of the worst GCT were people I had counted on to be there for me much of my life. Their inability to do so was another loss to my lifelong safety base.

The texts bombarded my phone.

GCT

Hey!!!

ME

Hey.

GCT

How are you?!!! What did you do today?!

ME

Read books.

WTF Just Happened?!

> **GCT**
> Oh, so you didn't take a walk today? Maybe you should go outside for a bit?!

The next day more exclamation-point-filled texts.

> **GCT**
> HEY!!!!

> **ME**
> Hey.

> **GCT**
> How are you???!!!!

> **ME**
> Not great.

> **GCT**
> Are you doing anything today? Going outside? Working?

I got it. That was exactly what THEY thought I SHOULD do.

Later, a family friend took me out to lunch. During the meal, work came up.

Me: I don't enjoy it now. I put my company on hold. I just… I don't know what I want to do anymore.

I was different and not sure what this new me looked like or what my goals now looked like. For someone who had always known exactly what I wanted, this was scary and disorienting. I would have loved to be able to have a raw and honest conversation about that.

Family Friend: You love work! You definitely need to be back at it.

Later, I got a text from a second Grief Conversion Therapist.

GCT2

Did you sleep last night?

ME

No

GCT2

Yoga helps with insomnia. Yoga and meditation.

ME

K.

GCT2

And work? Maybe get back to work?

Since my dad died, I felt a million times worse, isolated from who I had been and from a group of friends who loved to go out to all the newest spots and hook up with hot guys and obsess about their careers. I could not relate to them anymore. I could not relate to the me that could not relate to them.

Then there was my mom.

Mom: Honey. You look so sleepy. I just found out about this music that supposedly helps you calm your mind and fall asleep. If that doesn't work, maybe you should try sleeping pills. Have you been going to your therapist? Did you tell her you are not sleeping?

JESUS CHRIST! CAN'T IT JUST BE OKAY THAT I'M NOT OKAY! To make matters worse, the holidays were approaching.

Ever since I found out that Laura (or anyone else) could not download the consciousness of my dad at our upcoming Thanksgiving Dinner, I had no interest in celebrating.

His absence would hurt more than I could handle.

Mom: Honey, Eric and Kate are having some people over for Thanksgiving. They invited all of us.

WTF Just Happened?!

Eric is my mom's cousin. Kate is his daughter. They are wonderful people, but I had no desire to spend Thanksgiving with them. Eric's wife and Kate's mom passed away a few years earlier. You would think they would be a great support system. They weren't. They were major GCTs.

I got a text from Kate.

> **KATE**
> HEY!!! Are you coming this Thanksgiving???!! It will be so much fun!!

Yes. I am sure my first holiday season without my dad will be a total fucking blast.

(I didn't actually text that.)

I put down my phone and turned to my mom.

Me: Fuck Thanksgiving.
Mom: You might feel better and have a good time if you come celebrate.

I was so sick of this forced celebratory shit and everyone trying to "cheer me up." Somehow people seemed to think that by them acting extra happy around me and telling me to fake happiness would help me become happy. I wasn't sure if they even cared if I actually felt better as long as I acted like I did.

Me: YOU can do what you want. I will be in my room reading and gathering further evidence of an afterlife.

I checked my emails. There was a message from Phran.

To: Leigh + Liz
From: Phran
Subject: Thank you!
Hi Liz and Leigh.
Thanks for all you are doing! I assume you will both be taking a few days off for

Thanksgiving so enjoy the holidays and we can touch base after.
Love and misses.
Phran

I replied.

To: Phran
CC: Leigh
From: Liz
Subject: Re: Thank you!
Hey.
I hope you both have a great Thanksgiving and enjoy your time off. I don't celebrate holidays, so I will still be doing Forever Family stuff that day. Let me know if you need anything.
Xoxo
Liz

From Phran:

To: Liz
From: Phran
Subject: Re: Thank you!
Hi.
We don't celebrate either. If you want to join us and not celebrate you can come out.
Love and misses.
Phran

OKAY!

I confirmed with my mom. It was one thing to sit alone in my room and not celebrate, but it was another thing to not celebrate with another family.

She was fine with it, a bit relieved, too. I actually think she was scared I would bring up the afterlife to everyone at Thanksgiving. Instead of spending the day with my family, I took the train out to Phran and Bob's on Long Island to not celebrate. When I got there, we went into the dining room. There was turkey and wine, but no mention of holidays or Thanksgiving or forced happiness.

> **Me:** It's hard because my family wants to throw themselves in and celebrate and I don't want to. At all.
> **Phran:** That's normal. Everyone deals with their grief differently. It is a unique process. There is no right or wrong way.

We had a lovely night and I got to learn all about the start of Forever Family Foundation, early experiences they had had with mediums—some awe-inspiring, and some heartbreakingly horrifying—and a lot more about the scientists and other people in the afterlife research world. Through their stories, I got to learn more about who the mediums and other afterlife people actually were. Most truly were as kind, genuine, smart and thorough in their research as I had hoped.

And I got to ask Bob, my fellow skeptic, what I had wanted to know forever. What was his major turning point?

> **Bob:** One day Phran told me she heard this tapping sound and pattern she had never heard coming out of the car radio. She had a new car, so I didn't think anything of it. But it was interesting because it was happening even when the car was off. Then that same tapping which I realized sounded like morse code was happening in my car radio, even when it was off. Then we noticed it coming from other electronics around our home! We talked to Gary (Schwartz) about it, and he said for us to tell Bailey (their deceased daughter) to go talk to the medium Janet Mayer at a certain time. This was before we had ever met Janet and she had never even heard of us or Bailey. Gary told Janet to meditate at a certain time and then share what messages she got. After her meditation, Janet called Gary and said, 'I heard this weird kind of patterned tapping. It sounded like morse code.

What was that?' Then lastly, Gary was giving a big talk, we weren't there, but he called us and said he was trying to get ready for it and there was that patterned tapping coming out of the mic.
Me: Wow! So, unless Gary and Janet cheated...
Bob: Which they wouldn't. Just wait until you get to know Janet better. We have been working with her for over ten years now. She is as ethical as they come.
Me: No... it's just. Wow!

We continued not celebrating until about one in the morning.

Phran: I think this is past our bedtime. Aren't you exhausted? Let's show you to your room so you can get some sleep.
Me: I don't sleep. I have really bad insomnia.
Phran: Oh, okay. I had that for a long time. I'm glad to be done with that stage. It lasts for a while—at least it did for me. I'll show you to your room and you can just read or write or whatever you like to do.

I KEPT MULLING over Bob's "turning point." What would mine be? Had I had it with the "reading" I gave in Florida? Would there be something even stronger? What was the turning point for other skeptics?

I went into another workshop at IAC with that in mind. Because I was making up a previous class, I was the only one there, which meant I got the full attention of Michael.

Michael: From what I understand about you, you are much more science-minded than spiritual-minded. We can tailor this material to a more scientific perspective.
Me: Really?? Okay.

Once again, the relativity of time came up. For example,

someone could have an intense OBE that felt like a week, but when they came back, only a single minute had elapsed.

He taught me about different theories of dimensions.

Michael: Some dimensions have higher energy vibrations— this could be comparable to the buzz you feel in New York City versus a slower buzz you might feel out in the country. When you have OBEs, you could visit these different dimensions.
Me: They should do brain scans of people during OBEs versus dreaming versus normal life.
Michael: I agree. It's too bad there's not more money for that kind of research.

He explained about the energy of individual people.

Michael: Each person has their own energy: some are negative and some are more positive. We all can learn to sense that.
Me: This could also be helpful in everyday life. Being able to understand and sense energies like this.
Michael: Yes. It is. There are a whole variety of Energy Vampires. There are Sexual Vampires. And Intellectual Vampires—they just have to know everything. Emotional Vampires—they need you to be as angry or unhappy as they are.

Having people with bad energy explained and broken down this way was not only interesting, it was also freeing. I always felt guilty when I did not like someone for seemingly no reason. Now I wouldn't have to anymore. This explanation gave me even more permission to trust my gut and protect myself from toxic people. I could also use VELO techniques to clear out other people's energies and get in touch with my own, which would let me recognize if this "bad" energy was my own or if it were coming from someone else. Anyone could do the same thing.

Michael gave me more names to study in the fields of parapsychology and OBE: Charles Tart, Robert Monroe, Hal Puthoff, and Robert Crookall, among others.

Me: If there is so much strong scientific evidence of an afterlife and paranormal activity, why does mainstream science not even acknowledge this body of research?

What were they missing or more importantly what was I (and everyone else in this afterlife research world) missing?

Michael: In the culture of science, where everything has to be repeatable, that is not possible with this type of work. The scientific way of thinking developed during a very superstitious time when "witches" were burned at the stake. And there are still so many superstitions today, so it is crucial for science to stay clear and pure and not give into any hysteria, something this world of paranormal often does. Think about people who follow bloggers for medical advice over doctors.

Although according to Michael, his specific group of researchers, the scientists and doctors that he recommended are not prone to hysteria. They would study and examine abilities, recording the often-shocking data without jumping to conclusions. Other groups of researchers, doctors and bloggers behave like self-proclaimed gurus and are more inclined to rev up hysteria, drawing and claiming astounding conclusions for audiences, which should have only been possibilities.

Scientists, Michael continued, have to be very careful about protecting their reputation—and funding. If they were to publicly come out in favor of parapsychology, they could lose both. Scientists tend to be less emotional and more logical, but a lot of paranormal experiences require that people be more open to emotions and intuitions, like artists. At the same time, they still need to be grounded and focused on the objective data in front of them, not clouded by emotions.

So maybe none of those scientists could have that type of undeniable personal experience I had had in Florida?

Me: But, as Carl Sagan said, "Extraordinary claims require extraordinary evidence." Do you think there is an extraordinary level of evidence yet?

Michael: I have a different take. Parapsychology should follow the same standards as all other science. All scientific claims require SUFFICIENT evidence. In fact, there has been stronger evidence of many phenomena demonstrated in parapsychology than there has been regarding the efficacy of some approved medicines. The standards required for parapsychology are higher.

He and I practiced a few energy exercises—the VELO, moving energy in and out, feeling Michael's energy and the direction it was moving—all of which felt slightly stronger than the first time I tried. However slowly, I seemed to be getting more in tune with these energies.

Next, we moved onto a Remote Viewing exercise.

Michael took out a manila envelope. He told me there was a picture in it. Neither of us knew what the picture was. My job was to try to remote view into the envelope and draw whatever came to mind.

Without thinking, I drew a stick figure of a girl holding a suitcase near a stream. The actual picture was a cat in a party hat.

Then Michael took an object that he now looked at but I could not see, and went behind a wall. He thought about it and tried to send it to me. I tried not to think too hard. I said aloud and drew whatever popped into mind: an amusement park and fun, a roller coaster, a tree and sweets. An ice cream cone and an ice cream cup with sprinkles.

Michael: Do you want to see it?
Me: Yes.
Michael: A chocolate bar. Covered in nuts.
Me: Apparently, I am not much of a Remote Viewer and I guess I am not much of a mind reader either.
Michael: Are you kidding? You got it. Sweets—sprinkles. Amusement park, where you eat candy. Fun.
Me: I didn't get a chocolate bar.

Michael: It was the sense of what chocolate is—fun, sweets, the sensation of sweet food and candy. You drew ice cream and sprinkles. When you said that, I was like she got it! You got the sensations. Yes, there are people who would be a higher level who would get chocolate, but you got the concept.

I worried he was trying to make it fit, although not in a deceptive way. Maybe he was trying to get the theory across to me, giving me full credit for something I didn't fully grasp, like a tutor awarding a Gold Star to a struggling student.

Despite my initial skepticism, I also couldn't rule out that he was right, that maybe there was some psi going on. Weirdness was becoming a normal part of my life.

I couldn't wait to tell Phran about my IAC experiences. I also wanted to ask her some other questions about mediums. We had a call scheduled the next day anyway, so it was perfect timing.

Me: So, I had a few readings with these mediums who are not part of Forever Family Foundation. There is one that was good that maybe—
Phran: That one link you sent me? That one seemed like a crock of shit. Let me explain something to you, since you are so early in your grief. That one medium—I did not like his website because—
Me: But... they got like a million names of my family members accurate.
Phran: Listen. Don't interrupt. Just listen to me. Just because he got names right does not mean he was a good medium. Names can mean he is reading you psychically, not connecting with your loved ones. There are a lot of terrible mediums out there, and you need to be careful.

He had been my 14th reading. I had found him pretty early on, while exploring videos on YouTube. In one clip, he seemed to give a very evidential reading to a highly skeptical sitter. During this reading, the skeptic was blown away. He came out of the reading with a changed view of mediumship. After my own reading with him, I believed the video could be genuine.

> **Me:** Ohhhhh. I hadn't thought about that with the names.
> **Phran:** It doesn't necessarily mean he was, but did he get your loved one's personality? You are new to this so those are the things you need to learn to think about if you want to research this. I'm sorry, but suddenly everyone thinks they are a medium. That is one of the reasons we do the certification process with Forever Family, so mediums that are only psychic don't go around claiming they can do medium readings.

The next day, I felt extra sad. Probably all of it was just psychic. How could consciousness continue out of a brain? It made no sense.

That medium I just spoke about with Phran had told me my dad said to look out for my fridge. He said it wouldn't actually be broken but would go out for a while.

It hadn't. Just another example of how this was all bullshit. I started obsessing about the fridge and how it had not broken.

I was still staying at my parents' place. I went to open the fridge and it was not on.

What the fuck?

I went to the fuse box and flipped the switches. Nothing. The fridge did not turn back on.

Five hours later, it started working again for no reason.

Okay. That was weird!

Two days later, I was chopping vegetables on the counter by our sink, which has a light directly over it. I was thinking about how amazing that was, what had happened with the fridge.

Had my dad actually told the medium about the fridge and that he was planning to mess with it? Was my dad there with me when I was feeling sad the other day and knew I was thinking about how that medium mentioned the fridge and so he decided to show me he was around? Was it my own psychokinesis, the mind's ability to affect matter. Was it a coincidence? Another high-level coincidence.

Me: Okay, Dad… thanks for the fridge sign… if it was you. If this is all true, turn the light above the sink on and off.

I didn't expect anything.

Then the light very noticeably turned off for a total of about three seconds then turned back on.

I jumped back, threw the knife I was cutting up into the air and screamed out, "WHAT THE FUCK!!!"

This time there was no Zen reaction.

How did that just happen? Did it really?

A few hours later, I was sitting in my living room reading a book when suddenly in the middle of feeling pretty happy about the light, I was overwhelmed by this tsunami of aching and the unbearable pain of grief. Just when I started to think I was okay, I constantly discovered the many ways it would pummel me out of nowhere.

I missed my dad so much I felt I could not go on. I was furious this had even happened. How was this even possible that MY dad had died. How was the fact I could never see him so permanent and how was I so helpless to do a damn thing to change it?

In anger and pain, I threw my kindle down onto the couch and said aloud, "Fuck it. Can't I just cross already??"

This "grief wave" would have been followed by a burst of tears, had the living room lights not gone out—stayed out for a few seconds—then came back on. I was so shocked I froze, then started giggling.

Okay. Good one.

I had no idea anymore what was happening. My loved ones seemed to be learning how to reach out when I needed it the most.

Or I was noticing a lot of coincidences.

20

More Mediumship And Past Life Regressions

I was starting to notice I had some days where I didn't feel terrible all day. And some days I was even happy all day.

I am not saying it made my loss worth it, or that I wouldn't give up every extraordinary thing that happened and was going to happen in the future for one more day with my dad. I definitely would. And you probably would too. But still, some wonderful things did happen, and I was realizing it was not wrong to enjoy the wonderfulness of them.

I met some incredible people I would never have met otherwise. They were both changing my perspective and becoming people I cared about. I was also having so many transformative experiences that seemed to demonstrate the world was a more complex and more fascinating place than I could ever have imagined.

About four months after Florida, in March 2017, I built up the courage and started to try to "person" again. I packed up my stuff from my parents' place and returned to LA.

My mom had never lived alone in her entire life, and I was hit with a set of new terrors. If she died in her sleep, or fell in the shower, I would have no way of knowing aside from a few days of unanswered calls. I would then have no parents. I would be no one's favorite.

However, one thing excited me about returning, something that was bigger than my anxiety. I was taking a class with Renee Buck, the medium I had met in Florida. It was called "Developing Your Psychic and Mediumship Abilities."

We were assigned to read *Where Two Worlds Meet*[1] by a Forever Family Foundation medium Janet Nohavec. Janet is the founder of a church in New Jersey called The Journey Within—Spiritualists' National Union Church, which I later learned she preferred to call a spiritual center. By this point, I had gone there twice, watching Janet and her hand-selected mediums give highly accurate readings.

I had never before considered mediumship something that one developed. Aside from the fact I had thought it was bullshit until very recently, I had imagined that, if it was true, it was the type of thing that would have just happened to select people, something that would never have required work. Just how it is in the movies.

I walked into Renee's apartment. It was very Old Hollywood. The halls and staircases were covered in black-and-white photos of actors and actresses from the 1950s, which gave the building an almost noir-type feel. Her apartment was a sunny and spacious studio, with a shabby chic style, comfy plush chairs and a daybed filled with pillows.

Three other students were taking the class, including Jerome, the friend I dragged to Laura Lynne's event in Long Island.

On the couch seated among the pillows was a green stuffed rabbit. That was a sign I had set up with my dad! When I was little, he had given me a green stuffed rabbit. He would make up all these fun stories about it, inventing an entire personality for this rabbit, although its name never got more creative than 'Green Rabbit.'

Was this my dad letting me know he was around? Or was I becoming superstitious?

After class, I ran up to Renee.

Me: What was up with that green stuffed rabbit?!
Renee: Was that weird to have it out? My sister in-law made it for

me from my mother's robe, and I don't know why, but I was about to put it away. Then, for some reason, I decided to put it on the couch.
Me: No!! A green rabbit is a sign from my dad.

I had already had my reading with her.

Renee: Well, I guess my mother and your dad are working together to make sure you get your signs and evidence.

Each week, we worked on another chapter in Janet's book. Coming at it from a sciencey-researcher perspective, I was fascinated to observe this whole process from the inside.

We learned how to sort the information that came in if we were a medium giving readings. Make a chart in your minds-eye, like a family tree. Different family positions have different sections. At least that is how Janet and Renee do it. I assume each one has their own way.

With this tree chart, one side is reserved for the father's side of the family; the other is for the mother's side. For instance, if someone lost their grandmother, the discarnate would enter as a woman and stand two rows up on the tree. If the discarnate was their father's mother, they would be two rows up on the left side. If someone lost their mother, the discarnate would be one row up on the right side. Father, one row up on the left. A sibling or friend would be on the same level as the sitter. A child would appear a level below the sitter.

A medium required three things from the deceased person at the start of the reading—gender, relationship to the sitter, and age at the time of death. After that, they had to get a lot more information to be evidential, but those were the first three to identify.

Me: What would happen if you got someone who was transgender?
Renee: Perfect timing. That happened to me for the first time two weeks ago. A woman had lost her son who was originally born

female. I got both. I could not figure it out. I told her I was getting a child, but the child, an adult at the time of passing, was coming in as male and female.

As the class went on, and as I read more of Janet's book, it became clear that one of the problems with so many mediums is that they don't develop their skills to be highly factual and evidential. They think vague and loving messages are enough. Renee taught that your priorities as a medium should be to show, beyond a reasonable doubt, who you were "talking" to and that life does continue.

Yes! I could not agree more.

Apparently a reading with a medium is a three-way connection. If anyone's energy is off—the medium is sick or tired, or if the sitter is too close-minded or too skeptical—it could interfere with the connection like a radio with static. I had always assumed the sitter would not matter that much. That it was just between the medium and the person who had passed away.

Me: So, that seems like a catch-22. How would scientists then get evidence when they do tests and experiments?

Renee explained that you can still go in skeptical but not in an "arms folded and angry that this is a bunch of crap" mentality. The best way to investigate a medium was to go in as neutral as possible, keep an open- mind, and let the facts speak for themselves.

We sat in a circle and tried to get information about one another's deceased loved ones. We also brought in objects from our deceased loved ones and did various exercises, such as putting them into an envelope before anyone else could see them. Then we would all hold the envelopes and see what information came to us.

I wouldn't say I was highly specific in any information I got, but I apparently got a brother-in-law of Renee's. I described a man with dark brown hair and a mustache who lived in a cabin surrounded by hills. I also got some personality traits of her brother and her mother, but overall I didn't get very much.

In one class, we gave one another one-on-one readings. Renee rotated in as a reader and a sitter.

Me: You know too much about me now. This won't be evidential!
Renee: There's always more.

She then described a man who matched my great uncle who had passed fairly recently.

Renee: I see him as just a little under six-feet.
Me: Oh no. That part is wrong. I think he was about six-one or six-two.
Renee: Really? I dunno. He is insisting he is about an inch under. I feel pretty sure about that. Of course, I could be wrong.

After class, I called my mom.

Me: It was only a five-minute reading, but she described him physically and mentioned he was on your side a generation above you and got a little about his hobbies. But I know he was tall, and she kept insisting he was just under six-feet.
Mom: WHAT? Actually, Uncle Craig was five-eleven, one inch below six-feet, which made him so mad. He always complained about it. If any of this medium stuff were true, and I am not saying it is, he would bring that up, if he had to describe himself.
Me: Really? That also means Renee was not reading me psychically. You know how I have told you that is the dilemma that all of them, Doctor Julie Beischel and parapsychologists, come up against. This is great for the evidence of survival hypothesis versus psychic reading… you know… survival is where they are actually reading the consciousness of someone who has died, while in a psychic reading they are… you know… reading my mind and what I know or think about that person and then…
Mom: Okay! Okay! I do have to admit that is interesting.

During the next-to-last class of our workshop, things were

different. I had that "waves moving in my head, buzzy, vibrating" sensation I had experienced in Florida. But nowhere near as powerfully. I also noticed all my senses were heightened. Renee had candles burning, which smelled more potent than normal. The breeze from her window also felt stronger, more substantial.

One of the other students did a "reading," bringing in a relative of Renee's. Some of the information was accurate, some wasn't. During the reading, I suddenly got a terrible headache. I tried to ignore it. The other student could not determine how Renee's relative had passed away.

Renee finally told us the person who passed had brain cancer.

Me: They did!? I got this awful headache during the reading just now!
Renee: Oh, good.

She assured me in a calm voice that this was a normal thing to experience.

After class, Renee and I went on a hike. Everything was just *different*. Or more vibrant? I felt the air in a way I never had before. By that I mean, each subtle breeze felt stronger. I was more aware of the way air in California hiking in the hills in January feels different than air in, say, Paris or Thailand. Of course, I had known the climate was different, but the details of what makes one type of climate air different than another was more intense. And the smell from the flowers was more robust. My senses were gradually getting more sensitive overall, but it was always most noticeable right after class.

I told Renee.

Renee: Aww, that is bringing back what it was like for me when I was starting all this. That's normal.
Me: Sure. But I'm doing this from the Julie Beischel-esque science approach. Is this happening to her and those who study this all scientifically?

Renee: The further you go with all of this, the stronger your senses will keep getting.

It could be that all the meditations were growing more brain neurons, making me able to experience my senses more strongly. But I could also see how this possibly tied into experiencing mediumship. If our senses get stronger, and we experience a wider range of, I guess… the information waves… could they carry the energy from our loved ones? I remembered that was what the late medium Alex Tanous had thought, and why Loyd had had us do those exercises where we worked on increasing each sense. Five minutes a day of noticing another layer of details of sight, or sound or taste.

I did another thing the old me never would have, something that forced me to go even further away from my old sense of reality and further into the rabbit hole of weirdness: I had a past life / life-between-lives regression with Renee. Past-life regressions, according to Dr. Jim Tucker, Dr. Ian Stevenson, and afterlife science in general, cannot be considered evidence of reincarnation. The visions during the hypnosis of regression therapy can come from your own unconscious wishes, or fantasies. They also could have come from knowledge of certain eras, including knowledge you don't consciously know you have, that maybe you picked up in a book you briefly read in fifth grade or something. There is no way to prove otherwise. Once in a while, however, people have tried to trace details of their regression therapy to see if they matched up and had some luck. I don't know specific examples unfortunately.

Nevertheless, I wanted to try it.

Before the session, I had to fill out a form saying what I hoped to gain out of the experience.

Me: Evidence of an afterlife.
Renee: Aside from evidence. This will be questions about why you are who you are in this life. This means visiting three past lives. And some questions you would like to ask your guides.

I had to think a little bit, since really the only questions I wanted answered were more about data and facts. Did I actually live other lives? Together, Renee and I came up with the three (hypothetical) lives I would visit.

> **Me:** I guess I would like to know why I am still single? I mean beyond, 'Tinder sucks.'
> **Renee:** Good. Okay. We will visit a life that deals with relationships.
> **Me:** I guess one that explains how I ended up so easily finding the best people in the afterlife world. There is so much nonsense in this research. If that's what I had found first, I probably would have dismissed all of this.
> **Renee:** Okay, so we can look into a life where spirituality was a big part of your life.
> **Me:** And I would like to know my life right before this one.
> **Renee:** Yes. I always direct clients to go there. Sometimes, they don't land there. In that case, we will go with whatever life you happen to land on.

Renee explained that under hypnosis I would remember bits and pieces when I woke up. But I wouldn't remember all of it, just like what happens with dreaming. If it got too scary, or if horrible traumas came up, Renee would take me out of the intensity of it. I made her promise if I was an awful person, like some murderer or something, that she would remain my friend.

She promised.

I lay down on a bed and closed my eyes. She took me through some breathing exercises. I felt myself start to let go. It felt like I was floating, drifting off physically. My mind felt sharper than it normally does before I fall asleep, much sharper, in fact, but I also felt those odd waves again, as if waves of water were gently rolling through my brain and body. My brain felt light and sensitive, and mildly dizzy, like a little of that dizziness you get when you stand up too quickly.

But I felt present too.

She asked me to go back to a childhood memory. I went back to one of my favorite memories with my parents at our summer house. We were eating homemade ice-cream from the cute local little shop, sitting on a dock overlooking the lake. I always hated school, the exhausting competitive grind (even as a young kid), every day the same as the day before. Summers were definitely when I was happiest. I got to actually spend time doing things I enjoyed with people I enjoyed. My dad was telling us the story of the time he got himself kicked out of summer camp, because he punched the director's son in the nose! He was so mad about being forced to go to camp and miss out on a summer at his family lake house, he did what he could to get sent back home. I knew he had a temper, but it was odd to ever imagine him actually hitting a person. He laughed, but did not suggest I try the same thing. I wished I had had that kind of courage to escape the school I was in!

Renee guided me to feel everything, to taste and smell the ice-cream. I could, but it still felt as if I were doing all this in my imagination, although I did notice the more I focused, the more it started to become a stronger memory.

Accessing an early childhood memory was not hard. It was a kind of warm-up to then access a memory from our life we don't normally believe we can recall, before finally accessing a memory from a past-life.

After I described my family's lake house in detail, Renee told me she was going to bring me further, this time back to the womb.

My first thought: "What kind of weird person have I become?"

My second: "What the hell would Bill Nye think of me?"

And, finally, my third thought: "Considering all the weird shit you have experienced the past year, can't you just shut the fuck up and go with it!"

I listened to Renee's voice, focusing on counting backwards with her from 10. She told me she was taking me back to the womb.

I had a definite "Ewww" moment.

For the sake of evidence and experience, I pushed past the ick factor and imagined what being in a womb would be like. Warm, safe, dark... Then it got weird.

Some unexpected feeling took over my body. Factual and verbal adult concepts like 'this is what floating in liquid feels like' went away and sensations took over.

I felt myself tipping to the left. My body wasn't actually rolling or moving at all, but if I hadn't known that I would have thought the bed was tipping. I then felt myself roll slowly back towards the right before I gently tipped backwards. It was all dreamy and peaceful.

I suddenly felt an awareness of my arms and legs. They felt very short. Exactly how a babies' body would feel. It was all so illogical and so sensation-based and not how I would have consciously "imagined" being in a womb to feel.

The tipping feeling could be semi-compared to being drunk, where you feel as if the room is spinning but you know it is not. But my senses felt very sharp and I was serene, unlike when I am drunk with bed spins.

While most of me was serene, another part was still analyzing and observing, because this was really interesting! *I need to do an experiment with a machine that measures physiological responses as well as brain waves and see if responses match a person who was actually floating or tipping.*

I pushed that thought away. I didn't want to be analytical now. I just wanted to experience.

Renee: It is now time to be born.

As soon as Renee said that, I suddenly felt my body shoot upwards, faster and faster in the direction of the top of my head. Maybe it was the power of suggestion? Or could it actually be a deep muscle memory stored from when I was born? Either way, this was fascinating!

Then I was born.

The air felt cold and dry.

I looked around the hospital room. Everything was dulled down. It was

shadowy, as if it was dusk with no lights turned on. I wondered aloud why it would look like that when hospitals have such bright lights? Maybe it takes babies a few weeks to adjust to processing light and using their eyes?

I saw my parents' faces. They were huge. Was this what the world looked like through a newborn's eyes? Was I remembering?

Then it was time to float up and away and into a past life. Renee directed me to visualize going through a door, stepping into a past life. I pushed away my excitement and curiosity to focus.

Visualization was easy. I pictured a gilded door with an emerald green doorknob. Physically, I felt still all "wavy" and "vibrate-y" and far away from my normal physical world. The tactile sensations of turning the doorknob and feeling my bare feet in the grass were there, but the sense of touching was not as strong as visualizing. Still, touch was stronger than smell. I don't recall smelling anything.

After I walked through the door, Renee coached me to walk through the light and into my past life. I did. I saw nothing. I had no sense of time, so I am not sure how long this nothingness lasted. Finally, I saw a grey stone wall a few feet away. Beyond it were hills and a castle. Immediately, I knew I was in England during the 1700s. I don't know how I knew this; I just did. The facts of this life came quickly and clearly. As Renee guided me to share, the details poured out of my mouth without much thought. The visuals of my home, my clothing and the people around me were as vivid to me as if they were a memory from my own life. But I couldn't "remember" specific relationships or what was going on within these relationships. The physical sensations were still the weakest.

Renee told me to find a mirror and to look at my face. In this one, I was a girl probably in my late teens or early 20s. I had very white pale skin, a round face and very dark, almost black hair. Then more came to me: I knew I was in love, but the relationship didn't work out. It took place in England in the countryside, and I was apparently the daughter of a wealthy, titled family. I had fallen in love with a guy who worked at the castle. I tried to make it work, but all of the class barriers and cultural norms made it impossible. In the end, I settled and married someone I was "supposed to." The most interesting aspect was that I wanted this life/story to turn out one way—a love story with a happy ending, but I couldn't get it to. It ended with a bad heartbreak. I felt disappointment, but on the same level I would if I was watching a movie or reading a book. I was really emotionally engaged in the story, but I never felt the same level of emotion I do in my own life. I could, of course, have said that it had worked out, but something in my stomach felt off as if I was lying if I tried.

That was my life exploring a time a lesson in love was significant. I never really thought about how it would apply to my not being in a relationship yet in this life. I could have come up with reasons, but I was so much more interested in if this could be an actual past life or not.

In a hypnotic meditative voice, Renee guided me towards the next life.

> **Renee:** Now, we are going to ask your guides to bring you to a past life that sheds light on your spiritual journey in this lifetime from atheist to believer in the afterlife.
> **Me:** Scientific Explorer.

Even under hypnosis, I was not going to be called a believer!

> **Renee:** Okay. Scientific Explorer.
> **Me:** I wouldn't say believer. I seek evidence. I don't just believe things.
> **Renee:** Okay, evidence seeker. You are not a believer yet.

I wasn't going to let this go.

> **Me:** Believer to me means faith and I don't have faith or believe things. I absorb evidence. If the evidence proves otherwise, I change my mind.
> **Renee:** Okay. Very good.

I let the offense of being called a believer go, so I could move onto my next life.

The visual of the place poured in and took over. I felt as if I logically wanted to be in a city in Europe or America, but I kept seeing a gold sand street in a town. I was a middle-aged woman who had lost her husband. I was from Europe and had gone on a spiritual journey to somewhere either in the Middle East or the Mediterranean. Afterwards, I returned to Europe, where I also became involved in afterlife research. I visited different cities where I became involved in séances and exploration of the paranormal, while also sharing what I had learned overseas.

I saw my face vividly, but it did not look how I actually like to think of myself. I was much older than I liked to picture myself. I had dark brown hair as I do in this life, but the texture was different. It was wavier. I had slightly darker, more golden skin than I do in my real life. My features were larger.

All of this happened and unfolded in my head, and I didn't have much of a choice to edit it. It wasn't exactly like being told a story or shown a movie, but maybe somewhere in between that and writing it myself.

The last life I visited was my most recent. It started in the 1920s. I was a trendy "flapper" in my early 20s with light blonde, short wavy hair. My visit moved through The Great Depression and into the 1950s. It was not all happy, but not awful either. The Depression hit my family hard, as it did most people. What was interesting was I said I was concerned about my daughter in the 1950s. I worried that she was making overly safe decisions as an adult, because of how insecure her life had been growing up during The Depression. That was something that would not have ever crossed my mind. Renee told me that was apparently a common issue in the 1950s—people craved safety because of the terror of their early years in The Depression. I did not, consciously at least, remember hearing that before.

The final part of the regression was going up to a higher realm and meeting my spirit guides and deceased loved ones to ask them questions.

Renee: Close your inner eye and relax and let go of all you experienced from these lives. We are moving to the realm of The Other Side.

Suddenly, this strong breeze ran along my cheek. What in the fuck!?

Me: Is there a breeze? I felt a strong breeze along the right side of my face. Did you just blow on my face?
Renee: No. No, I didn't.
Me: There was no breeze?? Did you open a window? It was really strong.

Renee continued coaching me in a calm voice.

Renee: Maybe that was your guide zipping in close to you? Relax and close your inner eye.

She took me into a deeper meditation that led me through the door to The Other Side.

Renee: Let yourself soak in that warm pulsing energy. What do you see?

I never had told her about how I felt warm pulsing energy! Or that I was feeling it now!
I went through the light and stepped onto a cloud. Just when I told Renee that I was on this cloud, I felt a powerful breeze again move along my right cheek. It was even stronger than before.

Me: I felt a really strong breeze. I honestly did!

Renee continued to act as if this were completely normal.

Renee: Okay, so pay attention. Sometimes your guides join you. You are probably just feeling them.

Renee promised we could talk about it afterwards.

She posed some questions to my guides that I had sent to her in advance. The guides were to tell me the answers in my head, which I was to say out loud. The questions included some personal ones about my life, as well as bigger questions like why the world works the way it does.

As Renee asked each question one at a time, crystal clear answers that I had never thought of easily poured out of me. I had no emotions attached to the answers, unlike when I tried to think about these questions in real life.

One example was why had I not yet met my soulmate.

My "spirit guide answer" had a depth and calm insight I never had. It told me that I had more than one soulmate. We all do. I had not met him because I had been living on a lower level and hanging out with people on lower life levels. I had also spent a lot of time with people on different levels than who I truly was that were neither higher nor lower.

I understood these levels did not have the same shallow meaning such as money and social status with which we tend to associate "levels" in our human experience. It had more to do with levels of personal growth and our ability or desire to access a higher level or dimension. Living in the lower dimension and the wrong dimension for me had prevented me from even noticing my "soulmates." Now that I was moving more into who I really was and a higher dimension, my guide told me that I would start to meet more guys that would be more appropriate for who I truly was and who I was growing into. The official word mediums tend to use for this is "vibration."

This was all so clear. There was NO way I was not going to remember every bit. I knew Renee was recording this session and had told me that I would only remember scattered bits and pieces, but that didn't seem possible. It was all so vivid and clear during my past life "memories?" and guide questions that I was concerned I was not in a deep-enough altered state.

Renee had me say goodbye to my family and guides, then woke me up slowly while I kept my eyes closed and I stretched out my limbs. I was so disoriented that I hardly recognized the room. It was familiar but different, in the way that going back to an old place after some time and many new experiences feels, like when you visit your old college town.

I tried to remember the details from this regression, something I had thought would be easy when I was in the middle of it, but Renee was right. While I clearly remembered bits and pieces, I couldn't remember everything. The intense clarity I had during my visits vanished almost instantly when I woke up. I could only recall the visuals of my past homes and towns.

My conversations with my guides were the blurriest, which surprised me because they were so clear while they were happening. I now had only a vague memory of them.

> **Renee:** How long do you think you were under?
> **Me:** I don't know... about an hour-and-a-half?
> **Renee:** Four hours.
> **Me:** WHAT!? Really! And that breeze. Is there anything that could be?

We looked around. The windows were all shut tightly, and I was in the middle of the room, nowhere near anything that could have generated a sensation like a breeze.

> **Renee:** I was careful to turn the pages of the notebook I had of your questions to your guides so they made no noise. But, stay lying down. I'll turn them and let's see if you notice anything.

She turned them exactly the way she did sitting exactly where she had been sitting. She also turned them faster and closer. Nothing.

I took the notebook and waved it to make a breeze near my face, but it was a very different kind of breeze than the slim line of air that consistently ran along the center of my cheek, as if someone was blowing in my face.

I asked Renee if she had leaned into me at all. Maybe it was from her breathing, but she promised she had not moved from her upright position in the chair she was sitting in that was a few feet away.

I got up from the bed. I was a bit disoriented, but I also felt clear-headed and my senses remained heightened. I am not sure if I believe I actually communicated with guides and remembered past lives or if I accessed some inner creativity and a higher self, for lack of a better description.

As always with this research, the further I went, the more questions I had. If I had actually accessed my past lives, where was this memory bank stored? Did we access it more than we realized in terms of our daily thoughts and decisions and our personality in general?

While I was on the metro home my cousin texted me.

COUSIN
So, how was it?

ME
Interesting. You came up.

COUSIN
I did?

ME
Yes. Apparently, you and Donald Trump are lifelong soulmates, but in this life I came along since apparently he and I are eternal enemies, to be in your family and keep you away from him.

COUSIN
Fuck off.

ME
But it looks like in the next life you two will be reunited.

COUSIN
Seriously. Go fuck yourself.

21

The #2 WTF To Top All Wtf's That Ever Were

I had just gotten a text from Renee.

> **RENEE**
> Hey Liz. There will be three sitters tomorrow. Be prepared to give each one a 20-minute reading.

> **ME**
> No. No way.

This convo occurred the day before the eighth and final class I was taking with her.

When I arrived, the atmosphere was very meditation-conducive: candles burning, shades drawn, crystals laid out. I went and sat down on the couch to meditate. The room always had this very soothing and comforting energy, but it felt stronger today.

My senses were more potent. The smell in the apartment—a very nice candle that smelled like perfume—was more noticeable. I was more aware of the texture of the air and what it felt like on my skin. It felt a little bit like when you are feverish and hyper-aware of fabrics on your body or smells— this had that level of hyper-awareness minus the discomfort.

My head felt a bit warm and those waves were back, gently moving around me and in me. It was subtler than when I was in Florida, but it was definitely noticeable. As we began the meditation, I got a slight headache on the right side of my head. I had been getting those a lot more lately when I meditated or focused intently on energy.

Oddly, it wasn't a bad headache. It was almost like the way your muscles hurt when you get a deep massage on a tight knot. It was more of a pressure than a pain.

We finished the meditation and I was feeling a bit buzzy. My head or a limb would hurt for a minute as if this heat was kind of searing it but internally and pulsating. Still not in a bad way.

The three people we would be reading came in. I suddenly felt very- awkward. Renee pointed me to a chair. The first person came and sat in a chair facing me. She was a warm and kind-seeming woman, about my mom's age.

I began exactly how we were taught to:

"Hi. I'm Liz. What's your name?"

She introduced herself. I'll call her Katherine.

We were taught to behave as if we were doing a real reading and to explain how we get our information. For example, someone would say I am clairvoyant, then explain they see visually in their mind's eye someone who has passed away.

> **Me:** I am a medium. That means I can communicate with your loved ones on The Other Side. I am clairaudient. I hear what your loved ones are saying to me. I am also clairsentient, which means I feel them. Oh, and by the way, that isn't true. I have no abilities. We were just taught that is how mediums introduce themselves, so we are supposed to pretend we are mediums. Just so you do know I am not good or anything. I took this class more for a scientific curiosity.
>
> **Woman:** No worries. I don't expect anything.

I took a deep breath to relax and prepare myself for the start of a very boring and awkward 20-minutes. My brain felt warm. It was a bit like drinking a hot tea and feeling the liquid pour in and go down, but in my brain. Then I felt that heat around my arms too. I decided the best way to pass the first of the 20-minute sessions was just to guess a pile of bullshit and pour out every single random thing that popped into my head.

I started to babble, pouring out everything I was feeling (such as the heat) and say everything that came to mind totally uncensored.

I threw out something about a white dog, which I at first said was small. She corrected me and said she had a big white dog who died. I got a very vivid picture then of a fluffy big white dog with a pointy face going on lots of hikes. I described it. While I was picturing this white dog, I pictured a husky dog peeking around the corner with tons of energy. I pictured this dog grabbing food in a hyper-excited way. I pictured an orange cat like I once had, but a larger one that played outdoors.

I felt weird and kind of dreamy and caught into this story while I continued to say whatever came to mind. She kept saying most of it was true.

The woman was kind of there to me, but it felt a little like when you are telling a story from your life and reliving it while semi-aware of the person you are telling it to.

Woman: Yes, this is all true.
Me: WHAT? Wait, really? I just made it all up.

About five-minutes into this reading, Renee came over. She had been standing over me and listening the whole time.

Renee: It makes sense you would speak about animals, because Katherine (the woman) loves animals, but try to say something about people.
Woman: I have lost a lot of animals. Not so many people. I also connect more with animals than people.
Me: Sooo… umm… let me try to talk about people for the next fifteen minutes or so.
Renee: Actually, we are done. The twenty minutes are up.

Me: Huh!

The next reading was a guy about my age.

He sat down, and I introduced myself, just as I did for Katherine.

Me: So don't expect this to be good or anything.

I told him I was going to say anything that popped into my head uncensored.

My head was tingly, and I felt a pain on my right side and my arms felt warm. We were supposed to first figure out who their person on The Other Side was and feel and share their age of passing, gender, relationship and way they passed. I went to try to feel this. The first thing I thought of was a horse. Renee had just told me not to talk about animals, but try to get people. And my fucking brain gave me a horse.

I started to try to "feel out" a person who had died by moving my arms around on both sides of him, about six inches away from his body. I'm not sure why I was drawn to do this. It did not come from anything I had been taught. Maybe it was some instinct?? The area near the right side of his body felt warmer. The heat there felt stronger, thicker. I thought of the family trees Renee had taught us: father's side on the left; mother's side on the right.

Me: I feel an aunt or uncle that is in spirit.
Him: Yes, an aunt.

Then I kind of guessed, but it felt different than guessing. It felt the same as in Laura Lynne's class when I got 90 percent of the Zenar cards right and tested the difference between when information just popped in, versus thinking about it for a second. Here it was "popping in."

Try to do a brainteaser puzzle and you subtly can feel your brain working a bit. Think of a memory. Same thing. It's very subtle, but

THE #2 WTF TO TOP ALL WTF'S THAT EVER WERE

you can notice it. This was different because there was no feeling of my brain working at all.

I was very unaware of anyone or anything around me, except I was aware of what the sitter would say… very aware in fact. But I wasn't aware of them as an actual person the way I normally was when talking to someone.

I still felt the same sensations of warm buzzy waves which remained strong.

> **Me:** An aunt on your mother's side had two kids. About your age. A boy and a girl. But adults now. And there are little kids. Like between two and five. One of them has kids?
> **Him:** Wow. Yes.
> **Me:** Uh, was she a lawyer?
> **Him:** No. But my uncle, her husband, was a lawyer.
> **Me:** Hmmm—I'm not getting what she did?
> **Him:** Actually, a horse trainer.
> **ME:** SHUT UP!! I HONESTLY, AND YOU WON'T BELIEVE ME, BUT I HONESTLY WAS PICTURING A HORSE AT THE START.

He told me that sadly she apparently crossed in an accident while training a horse.

WHAT IN THE ACTUAL FUCK WAS GOING ON?! AGAIN.

> **Renee:** Okay, this was good. Try to stay in this stream of consciousness. When you get things right, you need to stop freaking out each time.

We then rotated to the last person. This person was a woman a little older than me. She seemed very patient. Despite the fact I had somehow gotten a ton of info right on the first two tries, I did not believe this was possible to do again, so I gave this woman the same spiel.

I took a deep breath and told myself I would hold in my amazement if I happened to get stuff right again.

I felt very drawn to her left side and, again, I reached out my arm and moved it alongside a few inches from her. I had no idea why I wanted to do that. My arm got very warm and the air felt thicker, and as if the buzz was faster, when I went to the middle part of her left side. I moved my arm above her to a parent and grandparent level or closer to below, a child level. When I moved my arm higher up to parent or lower to child the air felt a bit cooler and as if there was less movement or electricity. I felt compelled to keep it closer to her side and focus on the source of the buzzing heat.

> **Me:** Let's see… I… I feel more warmth or energy or something as if I am drawn to your left side, so I am going to say since it is on your left side, the male side. You lost a brother, husband or very good male friend.
> **Woman:** I lost my fiancé.

And as before I started to say a bunch of things I pictured in my mind. I honestly do not remember why I said this. If I felt a heat or just said it or what!

> **Me:** Did you also lose a friend? A childhood friend. A girl?
> **Woman:** Yes.

I again spewed out whatever came into my head. She said some apparently applied to her friend and some to her fiancé. She asked me which it was for but I did not know. I took Renee's advice and every time I got something right (which, once again, was a lot?!) I let myself feel the astonishment wash over me. I breathed in and told myself not to react to it.

After the shock washed over me, I was hit with a realization I was talking to a real person about a deeply meaningful relationship and loss of another very real person. This was not only a science experiment. This woman had lost her fiancé.

It suddenly felt overwhelming. What was startling, shocking and fun for me was sacred, traumatic and deeply meaningful to her.

After we were done with this last reading, it was the official verification part. All seven of us, the three "mediums," (one of the four of us was sick) the three sitters and Renee, sat together. Renee told them to evaluate how each reading went and what we got right. All of the sitters commented on how accurate I was.

Now was the time to freak out again.

Me: HOW DID I DO THIS? WHAT THE ACTUAL FUCK!
Renee: Liz calm down and let everyone finish.

They gave feedback to the other two "mediums." Jerome and the other student had given "life advice" and "loving messages from guides" but had not given accurate facts.

I was too disoriented and wavy to feel the full physiological effect of shock but I noticed my hands were shaking again. A LOT.

Jerome said he had heard me and noticed I talked a lot and very fast. He was surprised. I realized when he said that, the extent to which I had lost awareness of everyone else around me, because I had not noticed him at all.

Did I connect with the sitter into some altered state where I psychically read their minds? Maybe. Did I get into some hyper-aware state where I read every little cue they gave me on an unconscious level—from the tone of their voice to pupil dilation? I give this a maybe too, but unlikely since I was not looking at them and I was mostly visualizing things in my head. Also, that wouldn't give me the level of info I got? Or did I actually connect with their deceased loved ones? The laws of the universe and what was possible versus impossible kept drastically changing on me!

After "mediuming" at Renee's, I felt strange. Directly after the class, the waves stayed strong, and I felt an exhilaration. Whenever I lay down, it felt as if I was on a raft in a mildly wavy ocean. I slept well that night.

When I woke up, my body felt less like waves were going through it, but my head still felt those buzzy warm waves rocking my brain. I couldn't think clearly, but I still felt an excited high. By the middle of the week, it started to diminish, but the sensation was still there. I began to get worried—what if this never went away?

And towards the last few days, I was sick of it. It felt as if some strange energy kept changing the way my body was supposed to feel. Dr. Julie Beischel has noted that mediums had more health issues than the average person. Also, as many mediums told me, they are often exhausted after a reading, which is why they limit how many readings they do a week. They can also get dangerously sick if they do too many, especially with mediumship readings. Psychic readings are much less draining. I asked a few mediums why, but they are not sure why. If this was what they felt like, it could make sense why mediums have more health issues than the average person and could get sick from doing too many readings.

I also developed a very low-grade nausea as if I was slightly motion sick but it was also a slightly good feeling, similar to that sinking thrill when you go on a ride at an amusement park.

My brain still felt the buzzy, slightly headachy, hot waves that started at a spot in my skull and kept rolling through me, although a much lighter version of what I had felt during the reading.

After a week, my headaches finally went away completely. I was relieved because it was getting exhausting.

Mediums would say it was The Other Side communicating, skeptics would probably say it was psychosomatic or my brain still recovering from the trauma of my father's death. I had no idea what exactly it was, but I had no doubt there was something very powerful going on.

22

I'm Gonna Keep Trying This Out Of Body Thing

When I returned to New York for the summer, I went back to the IAC for another OBE workshop.

After the VELO warmups, we took out mats and lay down. They made the room cool, and we got under our blankets. They turned out the lights and flipped on a white noise machine. We began by moving the energy up and down and through us. Breathing deeply into my stomach was easier lying down than sitting up, but moving the energy was still just as hard.

I held the energy at my stomach. I experienced a weird combination of feeling panicky and irritable, as if I wanted to get up and run, and an acute sense that I was healing something that has been wrong for years. As I've said before, this is the main place where I've always held my stress. It was really interesting to think about how energy itself can loosen up those knots.

I slipped into a deeply relaxing, half-asleep mediation. I began to feel chills, the kind you get when you hear an emotional song. When I get chills in the normal world, they run along my body and build up to a certain point before fading away. Now, they built up higher than I normally experienced and stayed longer. I then began to feel and hear a subtle buzzing.

Robert Monroe, a "normal" and successful businessman had intense OBEs. He wrote about his experiences in his book, *Journeys Out of the Body*, describing how he would experience a loud buzzing before going out of his body.

The IAC told us the same thing. I felt as if all sensations extended out past my skin. I felt some powerful and hot energy for maybe 15-seconds at the bottom of my feet, as well as a few inches below them.

This buzzing and heat sensation went away, and I began to think about projecting energy out of me. Then I began to pull energy in, alternating between pulling in and pushing out.

It was as if someone was guiding me (at least in this dreamy state that is how I perceived it), to pull and hold energy in that tightness in my stomach instead of moving it up and down my body the way we were supposed to.

A message popped into my head that, until I loosened the energy blocks in my stomach and my throat, I would be limited in the amount of energy work I could experience, including having an OBE. I noticed some knots, or clutching, that I have had practically my whole life were slowly coming undone.

When we were woken up, I was disoriented and had no sense of how long I had been under. It turned out it was an hour-and-a-half.

We had time to ask questions.

I asked about my stomach and throat. The teacher Natalia said to push the energy between the two and try to push it out of me. Eventually, I will feel it shoot out. That sounded like it would feel amazing!

She also told us that we needed to trust our intuition. That sometimes we could have guides on The Other Side give us directions, but it would be in our nature to think we were making it up and we would ignore it. I have no idea where the message actually came from, but that was exactly what it felt like happened to me just now—knowing to hold the energy instead of moving it as the teacher had instructed us to.

I went back a week later.

The little buzzes and sparkles and warmth of energy felt like they were growing stronger the more I did my exercises, and I felt I had more control over moving it.

When it came time for our OBE exercises, Tricia, whose classes I had taken early on, took a seat at the front of the room and told us that she would be sitting in that chair the entire time. If we sensed someone moving around near us, that could mean we were out-of-body and sensing other consciousnesses also out-of-body.

We laid down and began one of the techniques. Slow, calm breathing with shorter breaths in and longer breaths out. I slowly felt my body relax. I worked on relaxing my stomach and breathing into it.

I felt a build-up of chills, just as I had last time. They didn't run through me and go away this time. They built up and stayed. Then my body awareness moved from my entire body to focus on only the front half, where I felt more chills and a dancing energy. These sensations stayed there and moved along and above my front, as if I had physical sensations and nerve endings up to a foot in front of my body. The sensations of these chills and energy then became stronger a foot above my body than on my actual physical body.

'Am I actually going out of my body!?' I thought as an intense excitement and then an instinctive fear came over me. Then it was as if these sensations and energy sucked back and absorbed instantly into my body. I was back in my body; the sensations ended at my skin.

GODDAMMIT!!

I went back to my breathing. The same above-body sensations returned. My head also felt that buzzy, waviness, which was quickly becoming a familiar sensation.

I then heard a slight buzzing and I saw, and kind of felt, these white lights sparkle and flash in the center of my forehead. There was a dinging sound that went off with them. This all lasted about five seconds.

I held myself back from jumping up and asking if someone had flashed lights over my forehead. But I had my eyes closed—and these were clear sparkling flashing bright white lights that could never have been that clear through my eyelids.

I calmed myself down and went into a deep, daydreamy state. I then saw a shadow out of the corner of my (closed) eyes and felt a presence. If Tricia had not told us that she would not be walking around, I would have definitely thought it was her. Or another student in the class. But there were only two other students in the class, both of whom had been lying next to me in a small room. I heard no footsteps. I had not heard anyone getting up. In the setup of this room you heard it if anyone even stirred.

Maybe this was the power of suggestion since she had mentioned this could happen?

Maybe it was shadow people as Stephen Hawking mentioned? That was what it must have been, I thought without the emotion and excitement, I have when these thoughts come to me.

The energy and wavy buzzy feeling were moving around me quickly now, and the sensations extending above my body felt more intense. I felt a little floaty. I breathed in and focused on the physical sensations to avoid getting fascinated and snapping out of it by starting to analyze.

My right arm suddenly felt extremely long and as if it was above my actual arm. This was similar to the way I kept feeling as if my nerve endings were a few inches beyond my skin. One technique to try to get out of body was to try to stretch this feeling of my arms all the way to the wall, without stretching my actual arms. I tried by feeling where the end of this "longer" version of my arm was, then stretched it further. It didn't work. So I lay there enjoying the intense warm buzzing energy until Tricia gently told us to start coming back.

I brought my full awareness back to my body, but the buzzing chills still felt intense. A wave of fear washed over me: What if I couldn't ever move again? The buzzing chills felt paralyzing, but in a way that had felt wonderful until this moment. I told myself that if I did have some unexpected reaction and couldn't move, I was surrounded by experts who would know what to do. My fear went away.

I moved my arms and they moved easily while that buzzy energy felt as if it was instantly absorbed into them along with the extended "nerve endings" and sensations.

I touched my hands to my head and bent my knees. Again, the chills and above sensations instantly absorbed into the rest my body. All sensations once again ended at my skin. I became aware of gravity anchoring me to the floor. My body felt very heavy and deeply relaxed.

> **Me:** Did you happen to flash white lights, like very sparkly ones over my face? And I know you said you weren't going to, but did you, or anyone in the class, walk along beside me?

No, Tricia had not flashed white lights over my forehead and no she had not walked around next to me. Nor had anyone else in the class. No one in the class had heard or sensed anyone walking around either. I believed her. I believed them.

> **Tricia:** Wow, Liz. It sounds like you were probably out of body.
> **Me:** REALLY?!!

I had no idea what to make of all this: A hypnotic power of suggestion? Hallucination?

The first steps of going OBE and connecting with other dimensions? The lights or sensations of some brain neuron reaction to that type of breathing? What was so interesting was that the times I did "medium readings," how I felt in Florida, these energy and OBE classes, and my dream of Laura and Lisa all had the same buzzy, rhythmic, wavy, warm sensations that also extended past my physical body.

Were these waves my key to other dimensions? To evidence—maybe even proof? To my dad?

23

Just Between Me And My Dad

One thing that bothered me was that my dad's favorite hobby, playing poker, never came up in my readings. However, I had gotten a few possible signs. One night, before going to bed, I asked my dad to send me one specifically about poker.

Playing poker had been part of our relationship when I was growing up.

Many nights when my mom was working late we played all sorts of card games, including poker. We would bet for candy when I was little, for money (no more than a few dollars) when I was older. Back then, winning two dollars felt like a big deal.

Once, when I was three, I was dropped off at a social club, where he was playing with his friends. Or so I was told. I don't remember. But my dad liked to tell me that I proudly announced to his friends that I beat him when we played. Obviously he had been letting me win.

He would meet a group of friends at a club in New York and play during the days. The fact he always managed his own work schedule (like my mom) was a value that meant something to me too. While both my parents worked hard and loved their jobs, they also loved spending time with their family and having hobbies they were passionate about. That was in contrast to many of the families I knew and the values I was taught in school. My dad loved poker so much he played it as much as he could. With poker being such a big part of his life, it made no sense that if he really were communicating, he had not "told" a medium about it.

The next day, I opened Laura Lynne's book, *The Light Between Us*, on my kindle that I was rereading. The chapter I happened to be up to was about a very skeptical scientist who got a mind-blowing sign.

I sent a message mentally to my dad that I wanted a sign of that level. And I wanted it to be about poker. Without giving it much thought, I clicked out of Laura's book and into another book by Dr. Gary Schwartz, which I had also been reading. The chapter I read mentioned a website that covered his work and the work of other afterlife researchers. I went to the website and clicked on a tab and was taken to a list of studies. I clicked on a random one. It was about two professional poker players. They were good friends and enjoyed playing together until one died. The surviving buddy joined an experiment with a medium who did not know poker well. (Scientists were involved to validate.) During the game, the medium channeled the deceased friend and played a few hands with the living friend. To everyone's surprise, the medium played at a professional level.

Thanks, Dad?

A week later I read another book by Dr. Schwartz, which included a chapter about a very skeptical man. He was a psychiatrist (like my Mom) and decided to attend a psychic workshop. During one exercise, everyone meditated then tried to receive a message from a deceased loved one of anyone else in the group.

The skeptical man said whatever came to his mind. He didn't expect anything. He didn't even think about what he saw. He just told the group that he saw two men playing poker. One of the men had had a special poker table that the other had accidentally broken. He owed him a new poker table. He said this message was for someone in the room called Kenneth. Same name as my dad! There happened to be a man in the group actually named Kenneth, who said he related to all of this. To say the skeptical psychiatrist was shocked would be an understatement. Relatable!

But why had no medium ever mentioned poker to me?

'I want to keep it just between us' popped into my mind.

I told myself I was rationalizing.

"Okay, Dad, if this is all true, then I want a direct sign about poker. I need more evidence."

The Forever Family Foundation has a weekly Radio Show. Their upcoming episode was for people to share how afterlife science had helped them. I told Phran and Bob that I wanted to share my story, and they gave me a 15-minute time frame to call in. I planned ahead so I would be home 15-minutes before my call-in time, and I would have been if the subway hadn't gotten stuck between stations.

The subway was stuck for 25 minutes. Finally, the subway moved, but to make my call on time I had to jump off at the next stop. I found a spot on a bench in Union Square and managed to call in just in time.

I briefly shared my story.

Despite the fact Bob promised me the mediums were not trolling the show to get information, I still did not share who I lost.

After the call, I was about to head back into the station to get onto my subway home, but it was a warm night, so I decided to walk part of the way home. I suddenly noticed that while lost in thought I was walking in the wrong direction. Not something I usually did in the city I had grown up in. Fuck it! I decided to continue on in the wrong direction to get to the nearest stop.

When I got there, I ran onto the train and grabbed a seat. I looked up and there was graffiti on the wall in front of me. The city's subways are very clean now, and I rarely saw graffiti on them. The graffiti right in front of me had the word "poker" painted with a pattern around it.

If so many little things hadn't been out of the norm, from a broken down subway to me walking the wrong way, and I hadn't happened to grab the exact seat I did, on this exact train, I never would have seen this (possible) sign. I sent a mental thank you as I held back crying. And this time I took a picture.

A few weeks later, I told this story to the medium Joe Perreta, the young one I had gotten to know in Florida. I told him how my dad's hobby had never come up even once in all of my private readings. I had had a reading with him so I could mention my dad, but I didn't tell him what his hobby was, in case it ever did come to Joe someday.

Joe P.: It sounds like he is saving that one to communicate directly with you.

24

Grief Retreat: Not So Weird Anymore

Forever Family Foundation was holding their annual summer Grief Retreat in the same bucolic retreat center in Connecticut they always did.

I joined a group of guests at one of the rustic picnic tables on the screened-in porch for breakfast. I found out I was seated with a group of parents who had lost children. I knew all of us at this retreat were hurting or we wouldn't be there, but I also understood that losing a child was its own kind of loss. I wanted to be mindful of their level of pain.

They were all lovely. I felt especially connected to one couple who had recently lost their adult son. They were both sciencey, too. One was a mathematician, the other was a heart surgeon. They had read and gotten comfort from the same books as I did—Dr. Jim Tucker. Dr. Julie Beischel. All the experiments at UVA.

After breakfast, I headed into a workshop with Gina Simone, one of the mediums I had sat with for a reading a few months earlier. (It was great.) I was happy to see the sciencey couple was in my group. We sat together in a small circle of about 15 people in an airy and sunny room with yellow walls, a floral couch, skylights and French doors that opened up to a large garden filled with grass, flowers and trees. Being a native New Yorker, it felt like being way out in the country.

Gina explained that we all have an electromagnetic field around us, the product of our own living tissue, which naturally generates energy. It's also how our loved ones communicate with us and it's in constant communication with our surroundings and with living people's energy. Because our loved ones vibrate at a higher frequency, we need to raise our frequency to match their vibration and connect with them.

> **Me:** What hertz do they vibrate at? Can't we create a machine that makes us vibrate at that hertz?
> **Gina:** It's not a hertz. It doesn't work like that.

She let us know that they also use dreams to communicate.

> **Me:** So, with dreams, why can't you ask them to tell you something you don't know but that someone else would know, then verify it and know the dream didn't just come from your own subconscious.

I had asked this before. No medium has given me a satisfactory answer. While this had not happened to me, I found out it HAD happened to Bob Ginsberg about another Forever Family Foundation member's loved one!

> **Gina:** You need to have some faith that it is them.
> **Me:** Why? Why would I ever have faith in that? Why wouldn't they do this for me, especially if they know me and know I would never have faith?
> **Gina:** I know you and I know you want proof, but you wouldn't demand God to offer you proof, would you?
> **Me:** No. No, I wouldn't, because I don't believe in God, so that would be a waste of my time. But, if any God ever wanted me to believe in them, I certainly would insist on valid proof before doing so.

It was interesting to notice that Gina, who I had seen give evidential readings and I knew how deeply and intelligently she thought about her gifts, turn to the concept of faith. I realized that most people raised with a religion are able to accept things that cannot be proven, something people raised without religion will usually never have.

She added that our deceased loved ones want us to have faith. That was why they won't come in too evidentially in our dreams. Other mediums have not given any better answers.

I couldn't believe our loved ones would demand faith in that way any more than a doctor who knew the treatments were working would withhold that information just so you could demonstrate faith in them. Demanding faith in that way was just plain sadistic.

And I said that.

I think probably the truth is that there are some, actually many, levels of science and laws of the universe that we are not yet capable of understanding.

When there aren't clear answers to something, especially if we are emotionally invested, we tend to turn back to what we were raised with. For Gina, this was faith, hence her answer. For me, it was that all this must just be a bunch of bullshit—hence my response that the only reason we couldn't get an answer in dreams was because survival was not true.

It very likely was neither. And the real reason was presumably as comprehendible to us now at our current level of scientific understanding as what would happen if we were to fly close to the speed of light in one direction for 10 billion years.

I came up to Gina in line for lunch right after this class.

Me: I hope I wasn't disrespectful! I would hate for you to think I don't respect your religion. And I hope I didn't make any religious grieving people feel terrible.
Gina: Not at all. It was great. You make the class think.

After lunch, Lynn Russel, a bereaved mother, gave a talk. She had written a book about her Near Death Experience (NDE) research. She had worked for Dr. Jeffrey Long, an NDE researcher I had heard of, although I had not yet read about his specific cases. I added him to my "to read" list. She said one main thing all NDEs had in common was that they were all full of love and joy. People usually met with their deceased loved ones during them. When people came back from this profound experience, they almost always said there was definitely an afterlife. Their outlook on life was transformed as well, convinced that the purpose of being on earth was to love. They also lost all fear of death; and money and success meant nothing compared to love.

> **Me:** Sooo... how do we know that they are actually meeting the individual consciousness of their deceased loved ones? Couldn't they just be seeing them because that is their definition of love? How do you know their loved ones are experiencing them back?
> **Lynn:** I guess we can't know for a fact, but there have been many evidential things that occurred, such as people who had NDEs coming back and reporting on things that happened in the hospital they couldn't have known. People who were blind described seeing and what they saw was verified. People mentioned conversations their family was having out in the waiting room. Do you know the story of Maria?

I knew about Maria. Maria was a woman who was declared dead after suffering a heart attack. She floated out of her body, then outside the walls of the hospital, where she saw a red shoe, including minute details such as a scuffed toe sitting on a window ledge. Maria was brought back from her NDE and shared her story with a nurse. The nurse went to go check if that shoe was really there. It was.

That was not an isolated case even if it was one of the more famous ones—much of what happens during these NDEs seems to check out in reality.

People have even met with deceased loved ones who had suddenly died while the person who had the NDE was in the hospital. The NDE person had not been told that the other person had died. Because they weren't elderly or sick and had died unexpectedly in accidents, they had not known that this loved one they encountered during their NDE was even dead.

I had read about many evidential cases like this in the works of Dr. Sam Parnia and Dr. Bruce Greyson, and cardiologist Pim van Lommel, all of whom are traditional medical doctors who investigate NDEs. I also read the work of psychology professor Dr. Kenneth Ring.

She then moved on, saying the universe is a sea of energy as demonstrated in the Global Consciousness Project.

> **Me:** So, I know the Global Consciousness Project started before the Internet was a thing. Has it changed since the Internet? It seems like consciousness spreads itself differently today than it used to. For example, my mom told me how everyone used to watch the same TV show at the same time. But she also meant just in the United States. Now everyone around the globe of certain generations watches the same things, but all scattered around the world and different times. And we watch a lot of things that are much shorter too—such as viral videos like Pizza Rat. Have patterns in the Global Consciousness Project responded to this?
> **Lynn:** I don't think that has been examined.

After the talk, that sciencey couple came up to me.

> **Woman:** You ask the best questions.
> **Man:** Yes. We get so happy when you raise your hand.
> **Me:** Really? I worry I annoy everyone.
> **Woman:** Not us. We like how you think about things.

I am not sure the rest of the group, or the mediums, agreed, but I loved that my mind could help others manage their grief.

That night, Marion Porter led a bonfire ceremony. Marion is a shaman and a good friend of Rebecca's. She invited each of us to throw in the bonfire a stick that we had blown our negative energy into.

The majority of people who went up together were couples who had lost a child. They stood holding the stick and holding one another, sharing a level of pain and understanding between the two of them that, despite the pain of my loss, was something I could not fully grasp.

Most people can't.

These couples held onto the stick longer than other guests. It was as if they still couldn't believe they were there and that this had actually happened. Maybe holding onto the stick could prevent the world from continuing without their child? These moments give me a perspective I had never before experienced. Yes, I did get that heaviness of grief, that thick, slow moving weighted-down energy that came into my body and settled in my stomach. But I also got that another layer of heaviness followed the loss of a child. I hoped the loving and fascinating energies that I was starting to feel were helping this couple—and every parent there—lighten their thick grief too.

After the bonfire, Rebecca grabbed me as I was heading into the end of the night wine and cheese event.

Rebecca: I know you hate smoking and are super into health, but come join me for a cigarette. I promise I'll blow the smoke away from you.

I followed.

Me: By the way, thanks for bringing up in your workshop that people feel suicidal. And that it is normal. Honestly, I would never have done it, but at the time I felt it and needed to share. But couldn't without people really freaking out.
Rebecca: It is part of grief. Why wouldn't you think of joining your loved ones? You know I meant what I said in that workshop, that I am jealous of you. That you never really believed any of this.
Me: Why would anyone ever be jealous of that? Before I had any hope of an afterlife as a possibility, everything was so bleak. I am still not positive this is all real and I wish I could be.

Rebecca: But I wish I could have known what it was like to have experienced that bleakness and a complete atheist view of the world.

I thought about how in the last week of my dad's life, I had started to read three books: One was fiction to escape and live someone else's life. One was *The Peaceful Pill*, which offered a comfortable exit from an unfathomably empty world. (Slitting my wrists seemed too violent and when I looked at my arms, they looked so much like arms, not just mine, that it felt as if I would be hurting another person.) The third was a book by Dr. Tucker, which gave me my very first glance into a world that ended up changing my life in ways I could never have imagined. As Phran once said, when confronted with this level of anguish, you either literally kill yourself or you figuratively kill yourself by not engaging in the world anymore. Or you get back into life and find ways to make meaning and help others who are going through the same things you have.

> **Me:** You know what. I think that I actually do get it. I could never have even fathomed an afterlife was possible, so the experience of constantly uncovering more and more... it's hard to put into words. It's so... astounding... in the best way. Almost like a magical treasure hunt. Then the comparison of my world before losing my dad versus after is not just a world with him versus one without him. It replaces the bleakness of loss with a certain magic.
> **Rebecca:** That is why I'm jealous. You have that comparison.
> **Me:** And you know what else? If I had always believed there was an afterlife, I would never have met any of you guys.

THE NEXT DAY Joe Perreta held his workshop. He shared an example of how he had developed his abilities. He had been working in a bar giving multiple 15-minute readings a night, putting in his 10,000 hours, the way you had to do with any other skill. He kept getting oranges.

He asked people he was reading if oranges meant anything to them. The response was a consistent no. He finally realized it might be a symbol for Florida. It was. Now when he sees an orange, he knows Florida is important in the person's life.

He then pointed to me.

Joe P.: I like to make sure I get my evidence. I'm like that one.

He closed the workshop with an hour of highly evidential and emotional readings. I ran up to Joe afterwards.

Me: That was amazing to watch. And I love that you question everything too.
Joe P.: Thanks. I would never just believe this. I need verification or I wouldn't do this job. But, I'm curious. Since you are so skeptical, how did you even decide to go to a medium in the first place?

I told him my story.

Me: I was lucky that the first medium I went to was highly evidential. But I always wonder if she was actually good since I never recorded it. The second one seemed honest, but she wasn't very evidential. The third was a total fake cold reader. I might have never pursued this world if I had the second one first and definitely not if I had had that third one first.
Joe P.: Who was this first medium?
Me: Oh, you don't know her. She is not with Forever Family Foundation or Windbridge or anything.
Joe P.: What's her name?

I told him.

Joe P.: I do know her! She is amazing. She is one of the best.

> Studies with Janet Nohavec. She's a perfect match for you.
> Believes a whole reading should be nothing but evidence.
> **Me:** Wow! It was total random luck that I even found her, and that she was the first one.
> **Joe P.:** Maybe it wasn't random or luck at all.

When we headed in for the closing ceremony, I noticed that the guests looked transformed. When most arrived, they had looked filled with despair. By the end, though, they no longer looked hopeless. Of course, they still looked heartbroken, but there was a new spark on top of that sadness.

Phran and Bob played the memorial video, which showed photos of the guests' loved ones who had passed away. The room became reverent and silent as everyone's losses now had a face. I am in charge of putting those videos together. I sit with each photo for a while and try to get who this person was. Is? When I meet the guests and they say names I can put a face, a real person, to their loved one. The final part of every retreat and conference is the same. The guests, volunteers, Phran and Bob and the mediums form a circle and then each person shares something they are going to take with them and something they are going to leave behind.

> **Me:** I am leaving behind another layer of skepticism. Well, skepticism is good, so a fundamentalist disbeliever type of skepticism, as Loyd described it to me. And I am taking with me another level of evidence, which gives me a hope I never thought possible.

A couple came up to me afterwards.

> **Man:** I also thought all this would be nonsense. I don't like flaky and woo at all.
> **Me:** I'm honestly still in shock it is turning out to not just be woo.
> **Woman:** I can't believe you never believed in God at all! And yet you still honestly really do think there is something to all this? Honestly?

I recognized the desperation in her voice.

Me: Yes. I promise. I really do or I wouldn't give my time to it.

They both hugged me goodbye.
I was teary on the train ride home. I texted my Mom.

Me: It sucks. We all bonded so much, and I might never see the other guests again and I won't see any of the mediums until our event in November. And most of the time they rotate mediums, so it could be years before I see these three again.

Then a text came through from a number I didn't recognize.

> UNKNOWN NUMBER:
> Hey, Social Media Girl! It's Rebecca.
>
> ME
> Rebecca!!
>
> REBECCA
> What are you doing in two weekends?
>
> ME
> Dunno?
>
> REBECCA
> Come to this Psychic Fair I am putting together and stay with me. You can meet Julez—my daughter. You would be friends with her. And Dusten Lyvers is coming and doing readings. Do you know him? He is also a Forever Family medium.
>
> ME
> Yes! I would love to meet Dusten and Julez and see you.
>
> REBECCA
> Great! See you in two weeks.

WTF Just Happened?!

Another nice surprise! Maybe I would actually get to have these people in my life more than once or twice a year.

25

How In The Actual Fuck??

There are certain things that happen along this crazy journey that are small drops in the bucket of evidence in possible favor of an afterlife. However, every now and then, comes a huge "the-odds-of-this-happening-by-coincidence-are-almost-none/turn-the-laws-of-science-upside-down" incident.

This was one of those incidents.

At the Grief Retreat, I had made videos interviewing the mediums for our social media, but the weekend was so busy that every time I tried to interview Gina something came up. When I attempted to find her after the closing ceremony, she had already left.

I ran up to Phran to tell her:

Phran: Do you need to be there to make the video or can you do this from the computer?
Me: No. I can do a FaceTime or something.
Phran: Okay, so email her a time to set that up.

On the train home, I shot an email off to Gina letting her know how badly I felt about leaving her out and we scheduled a time.

Solved. But, for some reason, I could not stop obsessing about it the entire ride. And this obsessing continued the next morning.

My mind kept repeating in a loop: *I can't believe I missed recording Gina. She is so good and that was so amazing when she got it right that my dad sent me the green feather.*

Let me back up a little bit. Very early on, after I was first learning about signs, I saw a large green feather on the street a block away from my parent's place. I assumed signs were nonsense. But a green feather, green my dad's color, an unexpected color for a feather, sitting right there. I picked it up. I would at least consider it.

Fast forward almost a year later when I had a reading with Gina. At the end of the reading, she asked, as all mediums do, if I had any questions for my dad, or my life in general. This is always a chance to talk more freely, not just answering yes, no, or maybe to a medium's questions. It's a bit more conversational. However, I did use it as an opportunity to get more evidence.

> **Me:** Has my dad sent me any signs?
> **Gina:** I don't like to say specific signs because you will be expecting one thing then disregard any other message you might be getting. Let's say you wanted me to say he sent you a green feather and...
> **Me:** Gina! It was a green feather.
> **Gina:** It was?! Well there you go. More evidence.

So back to the morning after getting home from the retreat, this obsessive thought-loop had grown much stronger. I had spent my first night home with my mom, and that morning I was heading back to my apartment in Brooklyn. The whole subway ride I kept hearing the sentence from my reading with Gina when she mentioned the green feather. The words got stronger the closer I got to my apartment. As I was heading up my stairs and then into my bedroom, it narrowed down further, and I just kept hearing "green feather." Those two words played over and over again in her voice, like a lyric from a song stuck in my head.

I grabbed my laptop to get to work and get on with my life. But this loop would not stop. I did my best to ignore it, but I could not stop mentally playing it. I then looked over and saw something out of the corner of my eye.

WHAT? Was I imagining this?

I walked over to get a closer look. A group of green feathers were sitting under my clothing rack in my room!? WHAT THE FUCK!

I examined them. Normal feathers?! I took a photo and I shot off an email with the photo attached to Phran.

To: Phran
From: Liz
Subject: WHAT IN THE ACTUAL FUCK??!!
Hi Phran.
So I promise I am not lying and I don't think I am hallucinating?? But I am not positive. Remember that green feather I got as a sign and then Gina got it right during my reading? Well... I was thinking about that all morning and then when I looked in the corner of my room there was a pile of green feathers.
Did they just apport or something? Is there a lab or somewhere I can send them to and see if they have any abnormal substance on them or if they are made of everything a normal feather is made of?
Can you look at the pic I attached and confirm you actually see them too?
Liz

I got an email back. Long story short, yes, she saw them. No, she didn't think I was lying or hallucinating. No, there wasn't a lab that could test them. What would they even test for? She let me know that that was a very special sign and that the further I went with all of this, she expected more and more amazing things to happen to me.

I do not own anything with green feathers. I have two roommates. Maybe one of them had someone over who had something with green feathers who then went into my room? Maybe one of my roommates had something with green feathers? If so, I had never seen it. I wanted to ask them, but I did not know how to explain why I was asking. And I definitely did not want to tell them the real reason I was asking.

Even if someone had stayed and they left them behind, that still could be a sign manipulated by my dad—or someone else?—the way my mind might have been manipulated when I stopped to look at the boat. The way that one medium said deceased people can use energy to press on our nervous system.

I had read and heard many stories lately where someone will feel illogically compelled to buy a gift for someone or post a photo of something only to find out that the item was a sign from the gift-receivers loved ones or a photo of a sign their best friend had just requested.

I counted the feathers. There were four. Since my lucky number is five I had a thought flash through my mind that I wished there were five. Then I felt bad because that seemed ungrateful, I mean come on—a group of green feathers that were my "sign" manifested seemingly out of thin air.

Two weeks later, on my living room shelf, I found one more, bringing the total number of feathers to five.

I have found no more since then.

26

Smoke and Mirrors and Séances

There is a small town, a village really, in upstate New York near Buffalo called Lily Dale, a community for Spiritualists dating back to the nineteenth century. Spiritualism is defined as being equal parts science, religion, and philosophy. They follow nine "guiding principles"[1] which affirm this definition. From what I learned; Lily Dale is a co-op village owned by all the residents. Only spiritualists are allowed to buy property in the village.

So, I told my mom and cousin that we were going to Lily Dale in the summer.

Mom: Not on your life.
Cousin: No way!
Me: We are!

I never got my Mom to cave, but after a lot of begging and harassing, my skeptical cousin finally, and begrudgingly, agreed to come.

The drive upstate was beautiful. Along the way, we stopped in charming towns with unique local shops and main streets with a lot of character that made us wish we had the time to hang out there.

We drove through winding hills with breathtaking views, and my cousin actually relaxed and became less pissed off that she was going to this "weird" place with me.

Until we arrived.

By the time we got there, it was almost midnight. We were about five- minutes away, driving along a bucolic country road with a lake on the side, when we came to a tiny church. A sign out front welcomed visitors to Lily Dale.

She had no problem making her feelings known.

Cousin: What is this shit? What cult thing did you drag me into?
Me: Shut up and give it a chance.
Cousin: Isn't that what people said about Trump after the election? Give him a chance?

We pulled up to a set of gates that marked the entrance to the village of Lily Dale. Because of the time of night, not a single person was out, other than the gate attendant, who handed us a paper map to guide us to our hotel. What was creepy was that the hotel had no physical address, so we couldn't enter it into our app. We had to figure out where we were, in the dark, based on a small paper map. We drove through the deserted, pitch-black village. The first thing we saw as we began to drive on the small dirt road into the community were two angel statues.

Cousin: You have got to be fucking kidding me. This is a literal church town.

I ignored her.

I thought the village was pretty magical. We were all alone, driving over semi-paved streets past rows and rows of Victorian-style homes. It felt as if we were stepping back in time.

Eventually, we found our way to the hotel, which would have taken us about one minute (Lily Dale is tiny!), if we had been able to enter an address into our phone. Cars were parked in front of a large white Victorian house with a wraparound porch and pillars.

Four women were sitting in rocking chairs on the porch. They asked me and my cousin if this was our first time in Lily Dale. They were guests at our hotel. The fact that they were staying up at night sitting on the porch and introducing themselves was the first indication of the special community atmosphere of Lily Dale.

It was early September in upstate New York. It was freezing outside and I was exhausted or I would have joined them.

We went back to the car to get our bags. I hoped the warmth and friendliness of the people would help relax my cousin.

Cousin: Those people were so weird.
Me: What are you talking about? They were friendly.
Cousin: This place is super creepy.

I ignored her as we carried our bags into the hotel. The lobby was just as historic as the porch. The floor was covered in green flowered carpet and the room was filled with comfy couches. There was an old-fashioned wooden desk with a bell to ring for check-in. A room to the side could be best described as a sitting room. It also had green floral carpet and was filled with white wicker furniture.

After checking in, we carried our bags one flight up a large carpeted staircase. I looked around the room. It had only two small beds with light blankets and no quilts. In the corner of the room was a plastic shower, which was identical to the ones we had had in our bunks at summer camp, and a small sink. Neither the shower, nor the sink were separated into a bathroom. My cousin watched me with an evil glint in her eye, happy to see me uncomfortable for the first time this trip. I didn't really mind the discomforts that much, though; it was a small trade-off for the historic charm of the hotel. I snuggled in as best I could and fell asleep.

The next morning, we both woke up early.

In the daylight, the village was even more of an experience of stepping back in time. The Victorian Houses looked like colorful dollhouses. Most had a wooden shingle out front with the owner's name after the title "medium." It was like entering a fantasy village from childhood.

There were three restaurants in the village, and all three were just as charming as the houses. Our favorite ended up being Lucy's. It was in a small house with couches and a swing on the porch. They had some delicious vegan pastries, and people were hanging out and socializing on the porch and at tables.

As if on a movie set, there was a little fire station and two town assembly halls—one was a small house and the other a large open-wall auditorium space. There were a few small chapels with stained glass windows, where healing and meditation sessions were held. Two general stores sold essential oils and crystals.

We entered the local one-room museum and history center. The older man who ran it was very knowledgeable about Lily Dale. He told us he had been visiting the village since he was a child. He showed us postcards from the 1950s, which showed young boys who were now middle-age Lily Dale mediums. There were photos of mediums from the late 1800s and early 1900s, old artifacts. My cousin lit up when she saw they had photos of Susan B. Anthony and the suffragettes. Apparently, Lily Dale was a center of the early women's rights and Suffragettes movement.

Me: Sooo… you are actually starting to enjoy yourself? Glad you went?
Cousin: Fine. I have to admit, this is all really interesting.

After my cousin bought a set of Susan B. Anthony postcards and properly Instagrammed the fact she had, we continued exploring. We walked around the little streets, bought some crystals and essential oils, and grabbed a bite at Lucy's.

Me: Whoa! It is already six. I have to get ready to go to the workshop for physical mediumship. Are you sure you don't want to come?
Cousin: Positive.

One of the main reasons I wanted to go on this trip in the first place was to witness a physical medium conduct a séance. A physical medium is different than a mental medium. Mental mediums like Laura and Renee were all I had seen so far. They connect with our deceased loved ones and tell us what they were "getting." Physical mediums can conjure physical phenomena, at least hypothetically. They make objects like jewelry apport (manifest out of nowhere), and some channel, meaning they let the spirits' voices speak directly through them. During these séances, the spirit voices have been reported to speak on their own, seemingly out of thin air. Physical mediums also call on spirits to visually appear, move objects around the room, and tip and shake tables.

Physical mediumship was a more-common occurrence in the late 1800s and 1900s. Not surprisingly, during closer examination, many were exposed as cheaters and frauds. A few I had heard of included a 19th-century physical medium named Daniel Dunglas Home, who supposedly floated in the air, and the well-known Fox sisters, who were said to have gotten spirits to communicate via tapping sounds during séances. Although there were lots of questions and doubts about whether any of this phenomena was real, especially coming from the Fox sisters. But it is harder to find people who claim such abilities today. This all sounded pretty hard to believe, but I had to see for myself.

I RAN TO THE CABIN, where the workshop was being held, about a five minute walk from our hotel. The cabin was rustic and smelled woodsy. Folding chairs were lined up in rows set for about 30 people. Most were filled. A young blond woman took a seat next to me and introduced herself as Leanna. She wanted to know what had brought me here. I gave her a brief rundown, about my loss, skepticism, and how now I wanted to check out physical mediumship. She shared she was there with her sister. They had lost their mom, although she was not at all skeptical.

Leanna: I think you will be amazed after this weekend!

The medium came in right on time. He was vibrant and dynamic.

Séance Medium: So why would someone choose to learn to be a physical medium? Why should you develop these abilities?

People began to call out: "To talk to spirit." "Love." "To communicate with ghosts."

Me: To gain valid evidence that we survive bodily death.
Séance Medium: No. There is one reason. To help people. That is it.

He kept the room engaged. He shared riveting, although hard-to-believe, stories. He definitely knew how to captivate an audience. Some claims he made seemed beyond outlandish, such as the "fact" that he had apported a missing ring that a woman had previously given her now- deceased daughter. To "prove" this really could happen, he showed a video of a medium apporting a jewel out of his eye. In the video, the jewel appeared out of nowhere, just as it would in a typical magic show, except supposedly with no illusion or sleight-of-hand.

Yes, this sounded preposterous. But, after everything I had seen, I still wanted to give him a chance.

He talked about ectoplasm, a material substance said to pour out of the bodies of physical mediums. This was frequently seen in the séances of the 1800s and 1900s and was almost always found to be stage magic. But there were a few times the researchers had not been able to figure out its source. The medium showed us a photo of it pouring out of his ear.

Me: So, what exactly is ectoplasm? Have any samples ever been analyzed in a lab?
Séance Medium: Yes. There is a study and they found it had three normal components: bleach, water, and urine, and three elements that are never-before-seen substances.

Me: Who conducted it?
Séance Medium: A lab in Switzerland.
Me: Which lab? And who were the researchers? And what year?
Séance Medium: Come find me after class and I will get you the details.

Leanna turned to me during a break.

Leanna: Ha, you really are sciencey! Wait until you see the séance tomorrow. That will blow you away.
Me: I hope so!

I was the only one in the class who asked any skeptical questions, although everyone in the room hung on his every word. He showed several videos of other physical mediums performing. One was of a famous "energy healer" I had not yet heard of called John of God from Brazil. In the clip, he pulled what looked like a toy chicken out of a large pot, which was somehow related to healing people. That part was not very convincing. But, just because a lot of what was in the videos was almost definitely fake, didn't mean they weren't exaggerated reenactments or entertaining representations of potentially real phenomena.

Séance Medium also talked about how we can feel subtle energies from one another. To demonstrate, he had us do an exercise. We all stood in groups and felt one another's energy, which was similar to some of the exercises I had done at IAC. Finally, he told us that physical mediumship was a skill that took years to develop.

I grabbed him after class to get details on the ectoplasm study.

Me: So that study? On ectoplasm. I would love to read it. Also, do you know of any scientists or researchers who are studying physical mediumship?

So far, all the researchers I knew of, such as Dr. Julie Beischel, were focused on mental mediumship. I wanted similar studies about physical mediums.

He directed me to the Society for Psychical Research (SPR), which, yes, I knew well. I was fully aware that they had not conducted studies on physical mediums in years. And as far as I knew, they had never analyzed ectoplasm. He also told me to check Basler Psi in Switzerland. Noted. I would. And in terms of the specific ectoplasm study he had mentioned?

> **Séance Medium:** I'll have to find all the details and send them to you. I think it was the SPR in the eighteen hundreds. Add me on FaceBook.

I broke my protocol and added a medium I had not had a reading with. It didn't matter that much for examining physical mediumship and I had nothing about my dad on Facebook anyway. The second I got back to my hotel I had a message from him.

> SÉANCE MEDIUM
> So, what brings you to Lily Dale?

I replied I was researching. That I took classes at The Rhine. And volunteered with Forever Family Foundation.

> **Cousin:** I think he thinks you're hot.
> **Me:** I think he wants to know who the person asking the sciencey questions is before they arrive at his séance.

THE FOLLOWING NIGHT, my cousin and I drove along a deserted country road to a small house surrounded by woods for the séance. Dusk was just settling in, and a chill was in the air, an almost ridiculously cinematic cliche atmosphere for a séance.

My cousin and I walked up a wooden porch and through a small coat room into a warm country kitchen. A large wooden table was set up. Cookies and chips were placed around it. I recognized a few people from the medium's workshop and a few guests from our hotel. There was an electrical excitement and anticipatory tension in the air, mixed with a tinge of the anxiety about the possibility of encountering the "supernatural."

As we filed into the séance room, my cousin drew me close, whispering.

Cousin: I actually am kind of nervous.
Me: Really?
Cousin: You have seen so much this past year. I… I guess I don't know what to expect.
Me: I honestly have no idea either.

Based on what I had read, I knew what *could* happen during a séance. What if a person's body manifested in front of me? Furniture moved? What if my dad's voice came through? This level of verification would turn everything I knew on its head. Even further than everything had already. I felt an exhilaration of possibility as I stepped into the room.

Inside, blackout sheets hung over the windows. A black curtain covered a door in the back of the room. Black sheets formed a small tent at the front of the room. The tent is called a cabinet; it's where the medium sits. Around the cabinet were approximately twenty folding chairs, arranged in a semi- circle.

We took our seats, and Séance Medium took his in the cabinet. I grabbed my cousin's arm in anticipation. She grabbed me back; I noticed her take in a deep breath. A collective energy of chills descended on the room.

Séance Medium explained the rules: No filming and no cameras, no drinking alcohol. No one was allowed to leave the room or get up under any circumstances once the séance begins. The room had to remain pitch black. His voice was grave. He let us know that this was all very serious and, while the spirits only come with love, if there was ectoplasm, any light would be very dangerous.

Séance Medium: In fact, four mediums actually died that way.

No, they hadn't. I had never read a single case where a medium died in a séance. What I had heard was when mediums insisted on pitch black, they were usually caught committing fraud, if lights were flicked on. Investigators usually insisted on infrared light, or in pre-infrared times, candlelight.

I was suspicious, but maybe he wanted to add to the atmosphere through a bit of creative license? I decided to go with it.

He paused, took a breath and, with the full tone and gravity teachers have when they know someone cheated on a test, scolded the group.

Séance Medium: I want everyone to know I had gotten a message from my spirits that someone was planning to secretly film this tonight.

A few audience members gasped. I am not sure if that was from the disbelief that someone would do something that had been defined as "forbidden," or because of the theatrics with which the medium presented it. Or both.

Me: So, what is the big deal about recording or taking photos?
Séance Medium: It is incredibly rude to secretly film and take content. I have someone approved to take photos when the spirits are okay with it.

He seemed angry that I asked. I made sure to let him know I got it.

He told us to leave the room so he could prepare. When we returned, he told us we would have to get checked to make sure we weren't hiding recording devices.

As we were walking out, he grabbed my arm.

Séance Medium: Not you.

OH FUCK. He really did think I was the one planning on filming.

I exchanged a worried look with my cousin, and the whole room stared curiously and nervously at me as they filed out.

I was taken aback by how scared I suddenly felt.

> **Séance Medium:** I was talking to my spirits about you and in the next twelve to fourteen months your research is going to get really big.
> **Me:** Really?
> **Séance Medium:** I want to work with you and I have some plans. It would involve travel and you could film me.
> **Me:** I am interested. I would love to do some controlled testing with you, too. I could bring in some researchers.
> **Séance Medium:** Let's talk more.

I stepped out. The room went silent; all eyes turned to me. My cousin was smirking. Without a doubt, she thought he had accused me of being the hidden-camera person, which I could tell she considered hilarious.

> **Woman 1:** So, what did he want?
> **Me:** Oh. I am super into parapsychology and he said he wanted me to do some research projects with him.
> **Woman 2:** What an honor.

Everyone was awed. I do not mean that arrogantly. Awe was the literal word to describe the way they talked about the medium and anyone who would be bestowed with his attention was incredibly lucky. I was excited at the possibility of researching him further and his possible openness, but I was also withholding any opinions until I saw him in action.

We lined up to go back in, and the anticipatory butterflies and static electric excitement in the air intensified. As we filed in, he checked the men, while a woman, who I guessed had been to many of his séances, checked the women. We had to leave our phones and bags in the main room, and they checked our glasses and eyeballs for filming devices, pockets for hidden phones, or who knows what else.

A woman next to us in line asked me and my cousin what brought us to the séance. I briefly but warmly told her that I had had a loss and now it was scientific curiosity.

Me: What brought you?

She had found this after a personal experience, which she didn't elaborate on, made her feel rejected and unloved by God. That caused her to look for other spiritual connections. Through these spirits, she told me, she learned God did truly love her.

Cousin: I am sorry you experienced that.

When it was our turn to be searched, I felt a momentary fear— would he plant something on me so he could publicly kick me out? No one here would have my back.

But it was fine, and my cousin and I and the rest of the guests all were cleared. I sat next to my cousin and grabbed her hand.

Séance Medium: Does anyone have any fear of the dark?

I wouldn't say I had an actual fear of the dark per se, but I did feel a bit nervous that being in pitch black could cause a weird sensory deprivation- type panic attack. I took a deep breath and reminded myself, if I was going to faint, I would just whisper to my cousin and, if it was serious, the séance would of course stop. The medium and his team would call a doctor.

The medium stood up and asked us to all take in one deep breath then say "Ohmmm" then take another breath and say "Ohmmm."

Séance Medium: I am going to need to move certain people around. The energy in the room needs to be just right for the spirits.

I was moved away from my cousin and into one of the front chairs near the cabinet. Now, what would I do if I was going to faint? Also, she was so skeptical. Would she have a panic attack if anything too weird happened?

Séance Medium let us know that the first two chairs closest to him on each side of the cabinet were reserved as a kind of VIP for his regulars. For some reason, I was seated in one of them. In the seat to my right was the woman who had searched us.

Everyone had known this was my first time and, once again, all eyes were on me. There was a feeling of *Who is this girl and how is she getting this treatment* from the ones who didn't know about my research.

Maybe he was trying to manage me and make sure I wasn't going to cause any problems. Or maybe it was because he knew I was connected to Phran and Bob and Loyd?

The woman seated on my left told me, as many others had tonight, that this would be absolutely life-changing. I felt another wave of thrill at the possibility.

A silence descended the room. Séance Medium explained that he would be gagged and bound to the chair. He had a cone, or a séance trumpet, which spirits supposedly spoke through. On the front of the cone were mini-lights and two drumsticks lay on the floor nearby covered with glow-in-the-dark tape.

He put on a sweatshirt inside out and put two pieces of tape in the shape of a cross on the front of his sweatshirt. He told us he would try to get the spirit to turn the sweatshirt right-side-out and put the tape back on in the same pattern. Considering he would be strapped to the chair that would not be possible by normal means. He asked two audience members to come up and check out the tent area and the armchair he would be strapped into.

"Liz, why don't you come up."

I was already at the edge of my seat wanting to investigate, but how weird that he again had singled me out. I felt around it. As far as I could tell, it was a normal chair, considering I had no training in stage magic.

We could inspect the curtains of the tent.

I did.

They were all normal, but the room was curved with a long wall along each side and, at the front of the room, a narrow area where we entered. The cabinet sat next to the entrance and, to its left, was a second door. This was where the black curtain hung. I pushed aside the curtain and gently pushed the door, which I knew opened into the coat room. I thought this was suspicious.

Now it was time for him to get strapped in.

He asked another audience member to come up and grab a closed bag of plastic lock straps. The audience member showed us that the bag was closed. He opened it and strapped the medium's arms to the chair. Before he was gagged, the medium said he would try to manifest ectoplasm tonight; he also promised his audience that we might witness a few levitations.

Levitations! The audience buzzed at the possibility. I certainly wanted to see that too.

He addressed skeptics' claims.

> **Séance Medium:** The skeptics will say I just move things with my hands, but you can all see how my hands are bound! They have even said I swallow bed sheets to produce this ectoplasm. Seriously. Now I want to find one person who could try that and survive.

The fact that any accusation of fraud was absurd was now planted in everyone's mind. The audience laughed along with him and rolled their eyes at the ridiculousness of the skeptics' claims. This made me wonder how open to testing he would really be.

Before we began, he asked who wanted to handle the opening prayer and who wanted to handle the closing. The atmosphere at this point was as if we were about to participate in a sacred ceremony. A woman took on the prayer role and said a spiritual and maybe Christian-based prayer that mentioned God being with us.

Finally, he was ready to be gagged and once we all saw him gagged, his words were muffled but clear enough for us to hear him order his assistants to close the curtain to the cabinet and turn out the lights.

The room was pitch black, but I could see the lit up tape on the objects and the light from the stereo. I could see nothing else in the room. I couldn't even see my own hand in front of my face. How long until phenomena began, if any phenomena would occur at all? The butterflies in my stomach grew. He asked the host to blast the music and the first song blared. From behind the curtain, we heard yelling and grumbling. Barking, he asked for the next song, then for the one after that to be skipped. "Next song," he said in a muffled, but distinct voice, "louder."

We heard more growling, feet stamping and yells coming from inside the curtains.

It was instantly obvious that this was a total load of bullshit. Classic and mediocre stage magic with all the distractions and misdirection. I felt a wave of disappointment, but I can't say I was surprised. I felt slightly embarrassed that I had built up such an anticipation. I switched my curiosity from one of paranormal interest to a curiosity to see if I could figure out how he was doing this.

His spirits that he apparently always works with came in. Each one had a unique personality. The first one who I will refer to here as Red Antelope, was a wise Native American. Not too long afterward, Winston Churchill came in. Yes, THE Winston Churchill chose to come to this small town séance, where not one person was related to him. A character I will refer to as "Tough Guy" who was scary on the surface, but loving under his rough exterior, showed up. There was a tinge of fear in the audience at "Tough Guy's" booming voice and gruff messages, but he always ended up making everyone shift to laughter.

All of these different character's voices sounded as if they came from the same man with the same accent. They were all coming from the same area up front. How was that not obvious?

The music came on again, louder this time. We were told to sing along. I kept quiet to try to hear any clues or cues about what was going on, but the people next to me were singing loudly. Did they actually still believe at this point?

FUCK! My cousin will never let me live this one down.

We had been warned that if we didn't have happy and engaged energy, the spirits wouldn't stay. While I could not see anyone, the room seemed dedicated to making sure the spirits stuck around. That involved loud singing, swaying and sometimes foot tapping. All convenient distractions from what was going on in the front of the room.

Red Antelope addressed selected members of the audience with messages of wisdom. First, the "spirit" addressed a man by name and offered him help on his growing mediumship. The man was truly honored.

Red Antelope gave a few more messages to people. Then his voice boomed out: "There is a Liz here."

Seriously?!

Me: Uh. Yes. Hi?
Red Antelope: I know all about your research and I want you to know it will be very important.

The woman next to me muttered "WOW" under her breath.

I didn't want to take away anyone's enjoyment. I tried to keep the amusement out of my voice.

Me: Thank you. That is very exciting.
Woman: That is truly amazing.
Me: Uh, oh yeah. Thanks. I'm excited.

I felt a little sad, and a little protective of the people who believed all of this. It really seemed to mean something special for them.

More music, more stamping of our feet.

Then Red Antelope said to prove that the medium was still in a trance the host should open the curtain and show us.

But not yet! (of course) After another song. After the song, the host asked: "Now?"

Red Antelope: No! In the count of five.

Oh, come on! This was beyond obvious.

Red Antelope dramatically counted up to five and asked the host to light a candle first, then open the curtains to the cabinet.

When the host opened them, the medium was "unconscious," gagged and bound.

The room gasped.

After the audience had a little time to digest the fact that the medium could not have been the one making the voices and movements, the host closed the curtain.

More music and noises and then a light show with four small lights that danced around. The trumpet floated and the spirit voices came out of it, one at a time. The finale of the light show was fast-moving drumsticks that beat in a vibrant and rhythmic dance. It was entertaining as a performance, but was very obviously a performance.

This wasn't entertaining stage magic to the people in the audience, though. To them, this was truly a sacred meeting of two dimensions.

After the light show, one of the, uh… "spirits" let us know that there was now a chair by the row of five squares of glow tape at the front of the room.

He directed us to count off in a circle from one to 22. Since it was so dark, it took some focus to realize when the person next to us and not the one next to them said their number. The "spirit" then said a few of us would be called by number to come up to the front chair.

He suggested the spirit would even levitate one of us.

There was this constant "maybe" to how far could all this go being dangled in front of us. I had to give Séance Medium credit: he knew how to captivate an audience.

Once again, I was one of the ones chosen to come to that front chair. This was getting ridiculous! Was he mocking me? Did he think his skills of illusion were so great he was going to win me over?

I carefully made my way, having to feel every step.

The music continued to blast. These "spirit" hands started rubbing my head. I was tempted to touch them or turn around, but that was against the rules, and despite knowing this was all fake, I was too curious to see how all this would play out to bother shaking it up. The hands were large and male and seemed to match how the medium's hand would likely feel, so I assumed it was him and not an assistant.

Needless to say, I was not levitated. (Nor was anyone else that night.) I returned to my seat guided through the pitch black by the voices of my seat mates.

The next person the medium called up was my cousin!? I could not wait to see how this played out.

I couldn't see her in the dark but luckily she was pretty quiet and did not call him out as he tried to levitate her. The "spirits" made a big deal about it, but then said it wasn't working. Oh god! What had she done to make them stop?!

A few more people went. More music. More showmanship. More distraction.

The "spirits" playfully bantered back and forth with some lowbrow jokes. This was getting stupider and stupider.

There was a little boy and girl "spirit" who were part of the medium's troupe. They showed up in the pitch-black room to play some games and scamper around. Despite being a young boy and girl, a grown man's voice spoke for them in a high-pitched childlike tone—with the same accent as the other spirits and, coincidentally, the same as Séance Medium himself.

It seemed that there were many regulars who had deep and personal attachments to all these spirits.

Then Red Antelope addressed another audience member by name. The woman replied, reverently.

Woman: Yes. Yes. Red Antelope.

Red Antelope: I know you want to be an energy healer.

Woman: Yes!

Her voice quivered with emotion, filled with hope.

Red Antelope: Keep working on it. Buy a blanket and when you lay it out you will be able to see inside your clients and you will know how to fix them.
Woman: Thank you. Thank you so much.

I did not laugh at that part.

What would happen when the woman bought that blanket and it didn't happen the way Red Antelope told her it would?

After a few more messages and music, the spirits said their goodbyes. The music blasted and, slowly, we heard the medium grunt and moan.

The lights came on, the closing prayer was said and the cabinet curtain was opened. Séance Medium was bound, gagged and barely conscious. The room responded in collective astonishment. This had been a transcendent experience for them, everything verified by seeing the medium bound and gagged.

The medium grunted and muffled by the gag managed to ask someone to remove it. Presenting himself in a still out of it semi-trance state he asked another person to check and then cut the plastic ties.

The person confirmed they were still intact. When they were removed, the medium showed us that his sweatshirt, just as promised, was turned right-side-out with the tape cross on it.

Leanna, the friendly and kind seeming blonde woman I had sat with at this medium's workshop the night before, came up to me and gave me a hug.

Judging by her glowing eyes, I could instantly tell what she thought of the night.

Leanna: See! You got your scientific evidence.
Me: Uh, yeah. Umm… lots to digest.

WTF Just Happened?!

We shuffled into the kitchen, and I looked around to gauge if anyone else knew this was a performance. However, every single person seemed to radiate a joyful glow.

Aside from my cousin, of course, who was thankfully politely hiding it. I walked over to her and whispered into her ear.

Me: Total bullshit.
Cousin: Thank god you said that!

The rest of the audience was eager to know how our first séance was, especially since I had so easily fallen into the coveted role of one of Séance Mediums' favorites.

I did not want to take anything away from anyone.

Me: Umm... wow. It was... well... beautiful. And fascinating.

Parts of it were.

I thanked the woman who sat next to me for helping me through my first séance. Before it started, she told me not to be nervous. She also told me that I should uncross my legs, or I would block some of the energies and my experience would be weaker. It had meant something to her to help out a newbie.

The woman who had been betrayed by God ran up to us.

Woman: What did you think?

She was radiant.

I hoped my cousin would remain discreet. The woman had already been betrayed by a God she believed in. She didn't need to experience further betrayal.

Me: It was really... phenomenal.

Thankfully, my cousin played along.
This woman shared more about her upbringing: she had been

raised fundamentalist Christian in a small town. She had become interested in parapsychology as a kid and the other kids, and even her family teased and shamed her for being a witch and devil worshipper.

I was glad that despite the bullying she had kept her curiosity in how the world could possibly work. She had been told the Bible had all the answers; to question or explore beyond that was considered wrong. Evil, in fact. It must be hard to remain inquisitive and continue to enjoy exploring and forming your own opinions when you are told so strongly not to. I had to admire her and felt a little happy that she didn't let this get sucked out of her.

Everyone was socializing in the kitchen enjoying the shared thrill and magic of the evening.

I wanted to get out of there; I grabbed my cousin. My cousin and I caught up in the car.

> **Cousin:** I was really afraid you thought it was all real and I wasn't sure what I was going to say to you!
> **Me:** Oh god, no. I knew the second the music came on. But I already had started to get suspicious when he created the hysteria about someone possibly filming. Just too much drama; it was so distracty. The real mediums have so much less hype and showmanship.

She explained what had happened during the levitation.

A few people—she didn't know which ones—shook her chair and seemed to want a reaction of amazement my cousin couldn't give. When she turned around, more out of curiosity than suspicion, she saw a man's arms quickly pull away. She didn't know if it was Séance Medium or an assistant.

> **Cousin:** I'm sorry. I know you are disappointed.
> **Me:** I am. A little bit.
> **Cousin:** I feel bad, though, because when I got back to my seat the people next to me kept asking in the most genuinely excited voices, "What happened?" I just couldn't play along so I told

them I saw arms. And they were like "Wow. Spirit arms? What were they like?" And I was like, "No. Human arms."
Me: Ohhh. Oh no. What did they say?
Cousin: Nothing. They just got quiet. And then it was as if it didn't register. People really can edit out or dismiss what they want to or what they need to.
Me: But it was not even an ounce believable.
Cousin: I know. Not believable at all.
Me: But the audience, they were all happy. I mean they were literally glowing.
Cousin: Yes. Yes, they were.
Me: And that fundamentalist Christian. She is going to believe something. This is better than Fundamentalist religion. Especially one that made her feel God hated her.
Cousin: Yes. It doesn't have the shaming and controlling that fundamentalist religion does.
Me: Is it that wrong what he is doing? I mean aside from the purity of science?
Cousin: Maybe not. I'm not sure.
Me: I'm not sure either. I honestly don't know what I think.

We got back to the hotel and a group was talking in the lobby. There were some familiar faces, including a few local mediums that I had joined in the lobby the previous night, plus a few new people. They stopped talking the second we walked in.

Local Medium: So how was the séance!
Another woman: We have been talking about it all night and couldn't wait to hear how it went!! Tell us everything.

What the hell was I going to say? My cousin and I froze. The pause was long enough for them to get it.
The excitement slowly drained from their faces.

Cousin: It was total stage magic.

They then turned to me hoping for a different answer.

Me: Uhh—well—I am not an actual parapsychologist. But I am into the studies of mediumship and all the research of the SPR.

I was babbling, nervously.

Me: And I couldn't see anything that could not have been done by normal means. I guess I would like to see this all studied further. Maybe have him do one under tight controls?
Local Medium: So why do you both think that?

I tried to explain as gently as possible, when a couple from the séance came bounding in. They were glowing, eager to report what they had seen.

Woman: Oh my gosh! It was absolutely incredible.

She rehashed the details of the night from a perspective of a world where the concept of illusion and deception did not exist.

THE NIGHT before we left Lily Dale, I returned to the séance. This time on my own. Afterward, I filled in my cousin about what happened, during our final walk around the fairytale-esque village.

Cousin: Did Séance Medium suddenly show magical abilities on the second night?
Me: Aside from a magical ability to mentally manipulate people using overly dramatic stage magic? No. But it was so weird. He picked me as a favorite again.
Cousin: I told you he thinks you're hot.
Me: Fuck off. He has little kids and I think a wife or girlfriend.
Cousin: Ahhh, then no. He is so ethical. I am sure he would never cheat.

Me: Point taken. But seriously. It was strange. For some reason, he chose me to do the body pat down for the women. I felt awkward and a little invasive.
Cousin: Seriously? That is hilarious.
Me: And Winston Churchill came back and mentioned my research again. He said, if it got good enough, I could test Séance Medium, even draw his blood, in a year or two.
Cousin: Yeah, I'm sure Séance Medium will allow that! But can I name drop that Winston Churchill thinks your parapsychological studies are impressive?
Me: Ha! Go ahead. But the craziest part was when he manifested ectoplasm and…
Cousin: What do you mean? Obviously that was fake.
Me: Of course. Let me finish. Red Antelope claimed there was ectoplasm coming when he was behind the curtains. It was a huge dramatic buildup. I was honestly a bit excited. I know this sounds silly, but so many historic parapsychologists have caught mediums using fake ectoplasm that I felt as if I was part of parapsychological history. That I was about to witness an ectoplasm stunt.
Me: The curtain opened, and the room gasped. We could all hardly see, but it looked like this white kind of fuzzy material hanging from his face. The music turned off and everyone was whispering in awe and struggling to see it clearly in the dim candlelight. But then he called ME up to the front to sit and get a closer look. Me! The one who would clearly realize it was a pile of cotton, to come up and see that was exactly what it was, a very obvious pile of cotton.
Cousin: What did you do? I would have started laughing.
Me: I was so baffled at why he was letting me get a closer look. I stared at him, smirking at the absurdity, but he prompted me to say if I saw it. I played up about how incredible it was and what an honor it was in a voice that was sarcastic enough that I think he got it but not so sarcastic that everyone else got it.
Cousin: What the fuck! Why do you think he did that?
Me: I have no idea. Oh, after the séance I snuck in and found his

suitcase with the props. I didn't dig too deeply, but I saw the ectoplasm—it looked like a huge Santa beard.

Cousin: That is hilarious!

Me: There's more. Before heading to the séance, I ran into Leanna, who wasn't going to be able to make it to this second night of the seance. But she was so excited. That morning she saw a chalk print of a "spirit's hand" on her car seat. I couldn't figure out what that could be—if she was way exaggerating or if somehow someone had put it on her car. Then, as soon as I walked into the séance, Séance Medium was like, "Hey, Liz, so you want evidence, guess what? Remember yesterday when I told Leanna a child spirit was going to go home with her? She saw a hand print in her car!" Can you believe that!

Cousin: What the fuck!! He's so obvious. But how sad to play with people's wishes and need to believe. It's almost… violatey? Did you tell him not to break into cars?

Me: Yeah. True. He's really sleazy and so absurdly transparent. But I'm still not sure if it was him or an assistant. This also explains one of the reasons we're not allowed to bring our stuff in, so he can take our car keys or put crystals into our bags. I spent the whole séance thinking, *please don't try to do something to my stuff and lose my car keys or anything.*

Cousin: He is not even subtle! But why do you think he kept selecting you? For the ectoplasm? For all of it?

Me: I'm not sure. I wonder if he was just bored with everyone always being so believey and wanted to push the envelope a bit? Or maybe he wanted to test me to see how much I would play along so he could plan his next move since he knows I'm writing.

Cousin: Maybe?

Me: I tried to investigate further. You know how I told you I felt the door in the cabinet and I knew it went into the coat room where we left our stuff? I left my bag against it to see if it opened during the séance. I put a piece of tape over where the door and the wall met. I was going to grab a photo but then the owner of the house came in. I'm sure she would not have appreciated me trying to catch Séance Medium cheating. I just said the first thing that

came to mind—she had these kids drawings on the walls so I asked if they were done by spirits.
Cousin: Wait, what? Seriously??
Me: It was kind of a panic reaction. I wanted her to think I was super believey so she didn't notice all the traps and stuff I was trying to do.
Cousin: What did she even say to that?
Me: I think that went a little far for her because she looked at me as if I was a complete idiot. She politely said "human child."
Cousin: OMG. That is so funny!
Me: Anyway, the tape didn't move, as far as I could tell.
Cousin: Ha! You would really ideally have someone standing by it.
Me: I know! Or a recording device. But, as I was saying, I'm still trying to figure out why he kept choosing me.
Cousin: Maybe he is so used to believey people he thinks his tricks are much better than they are? Or he wanted you as the skeptical one to verify it for his people? I can't deny his psychological games were brilliant manipulations. The suspense. The us-versus-them of the skeptics and the "someone is secretly filming," all making people feel a need to please him and prove their loyalty. That kept them busy proving their loyalty in their own minds too.
Me: Oh, he pulled that last night too. 'Red Antelope' interrupted the séance to say he got a warning that someone was planning to film and…

We were so lost in conversation we almost bumped into Leanna, who was on her way to grab lunch at Lucy's Cafe.

Leanna: How was the séance last night? I am so jealous you got to go a second time.
Me: I don't know. I am intrigued enough to want to study this further with scientific controls.

I did not have the heart for anything else.

Leanna: I think the further you get, the more proof you will get. Did he have ectoplasm last night? Did you get to see it?
Me: Actually. Yes. He did. I did.

Leanna's face lit up.

Leanna: What was it like?! I have always wanted to see it.
Me: It was so dark; it was hard to see clearly from my seat. Uh but I… I saw the outline and shadow of it.
Leanna: I cannot believe you got to see ectoplasm. You are so lucky! I get it. I was a little skeptical at first too. But now you actually saw ectoplasm. That is proof.
Me: Uh, yes. I uh… I guess so.

After we hugged goodbye, my cousin and I settled back into our seats at Lucy's.

Cousin: Leanna was sweet. I did end up liking her. And everyone we met. I just can't take how easily they believe things. Believing something because they are told to just bugs me. It's so dangerous. It's why people are not vaccinating kids and the measles are coming back. It's why Trump won.
Me: He didn't win, though. He got fewer votes by far. The electoral college is why he won.
Cousin: But he should have gotten something like thirty percent.
Me: He should have gotten three percent.
Cousin: True.
Me: I can't figure out how much they believe versus how much they make themselves believe. This one kind-hearted woman I sat next to last night kept trying to convince me. I guess she could somehow tell I didn't believe. She told me the thing that pushed her over the edge into belief was that the characters were all so different that there was no way he could be such different people.
Cousin: I don't think it was you she was trying to convince. It also bugged me because it was so silly and overdramatic.
Me: True. One of the things I have come to appreciate is that

there is a subtle beauty to everything I have encountered that was probably genuine. This performance was so blatant and crude. Real abilities are never like that. His real abilities, if he even has any, would never be this entertaining and they would never be this reliable. These abilities, if ones like these could even exist, would be very subtle, like the Scole Experiments or things Loyd mentioned in *Mind Over Matter.* We might sit in semi-dark for three hours and just a few lights would move, or a folded paper on a pin could be made to spin. We would feel a cool breeze like I felt with Renee. Maybe a voice would say one name. But, unfortunately, I don't think most people would pay for the real stuff.

Cousin: It's like reality TV culture.

Me: In fact, Loyd taught us that originally physical mediums would work in the light but when séances and physical mediumship became popular—I think this was in the late eighteen hundreds—fake mediums popped up like an epidemic and would do their work in the dark for obvious reasons. When the genuine ones would hold their séances in the light, everyone thought they were the fake ones.

Cousin: That is interesting.

Me: So… what if he has some real abilities and wants to share the hope and knowledge with people but his body can't do it that much. He knows this isn't a scientific group so he isn't going to waste his energy. He might as well save it for an experiment with parapsychologists. And maybe that is why he revealed the ectoplasm secret to me. So I could see he was admitting he knew these performances were absurd.

Cousin: Hmmm.

Me: He could be doing these to show this is all real in a way people can digest. And he wants to make money off of his skill. There's nothing wrong with that. So he isn't lying or deceiving by faking a highly dramatized version of what he can actually do.

Cousin: That is worth investigating. If people are going to have blind faith in something, this seems like one of the less harmful things to have blind faith in.

I didn't think anyone I had met was a fool. Believing this was all true did not seem to be a matter of intelligence as much as a choice (albeit probably unconscious) that they were going to turn off the part of themselves that would think critically.

I had to agree with my cousin that religion seemed to be a part of it. Whether or not they still followed a traditional religion—most of them were raised Christian—it seemed they were all comfortable with being told something was true and just accepting that it was.

I am sure there are communities of all religions that encourage deep thought and questioning, but many of the people I had met seemed to have come from that "because I said so and don't question" kind of religion. And maybe education too.

Also, the wish for something to be true can do a lot to blind people from the obvious. Everyone who has lost someone to illness thought for at least a period of time that their loved one was going to get better, despite all evidence to the contrary. And almost everyone has had a relationship they knew wasn't going to work out, but stayed in it anyway, telling themselves it would.

I have done both of those.

> **Me:** You don't think that I have been fooled the way everyone at the séance was by my mediums and research?
> **Cousin:** No. Not at all. I trust your judgment. From what you have told me about your mediums and the things you have experienced, they all seem to have real substance. They are even starting to change my mind. Plus, you knew the séance was bullshit the second it started. And, when you had some bad medium readings, even at your most desperate, you never lied to yourself.
> **Me:** The thing I have noticed in studying all of this was that there wasn't one dramatic bang that transformed everything in one evening, the way this séance pretended to. It was all subtle and profound, although often still spine-tinglingly wondrous. It has a beauty this séance lacked.
> **Cousin:** That makes sense. What I'm most interested in figuring out is how it is that so many people just relinquish their ability to think about something critically.

Me: And, honestly, you promise you don't think I'm doing that?

I had to ask again.

Cousin: No. I honestly don't. You still think about all your experiences from many angles. And you also say you don't fully know what any of this means. Saying you now think an afterlife is possible or even probable isn't the same as just believing anything. You have gotten valid evidence of some amazing things. Plus, didn't you really want this event to be true?
Me: Yes. Yes. I really did.

27

Told-You-So's and Secrets Revealed

A week later, I was at lunch with my mom, filling her in on the fraud of Séance Medium.

Mom: Sounds fascinating. What a mind-manipulation that guy did. You will have to fill me in on the rest later, I have to hurry because I have to read a case for my group.

My mom, the successful psychiatrist who was more skeptical than even my cousin, was part of a group of psychiatrists and psychoanalysts who read and analyzed various case studies of consciousness, the brain, and individual patients. They met once a week to discuss these cases.

Me: Oh, tell me about it?

She took a deep breath.

Mom: Well… okay, you are going to LOVE this. We are reading a case by Jim Tucker this week, then another by Ian Stevenson next week.

Me: WHAAAAT!!!
Mom: Yes. Yes. I know.
Me: Which cases?? Which!
Mom: This one is about a boy remembering being a pilot or something.
Me: James Leininger? The one who has memories of being a World War II pilot in a past life?

This is the most mind-blowing case ever. Apparently, it was the case that convinced Dr. Jim Tucker that we do live multiple lives, although he won't publicly conclude. At three-years-old, James Leininger remembers specifics of being a pilot during WWII, knows details about the plane, names of his army troupe, and details of how he died when his plane crashed. The details check out. He had soldier dolls, each with a different hair color. He had named each, and Dr. Tucker later found out the hair color and names corresponded to hair colors and names of his army buddies. He meets them when they are old men, while he is a little boy in this new life, and recognizes some of them and shares memories with them. He meets with his now-elderly sister from his past life and shares enough memories that she is convinced. There is much more to it, and Dr. Tucker goes into the details in full in his book.

Mom: Yes. I think that is the one.
Me: And Ian Stevenson? Which cases?
Mom: I think it is about kids who have past life memories and corresponding birthmarks to match.
Me: Call me right after your meeting!

Later that night. Phone rings.

Me: Sooooo...
Mom: So, I have to admit it is pretty amazing. But there are many things to consider such as unconscious fantasy and repressed wishes. Living out the repressed wishes of the parents. The unconscious is powerful.

Me: Yes. Yes, I know. And Jim Tucker addressed that. There is no way that could explain what happened in this case.
Mom: Okay, true.
Me: And what did everyone else think?
Mom: Well, I have to admit your Jim Tucker makes a great case.

The next week we resumed our conversation.

Mom: So, we finished Ian Stevenson. About the birthmarks.

In the birthmark cases, kids will have memories of past-life traumatic deaths and then will have birthmarks, which match the injury of their death. For example, if a kid remembers being shot in the back of the head in a past life, Dr. Ian Stevenson has found a birthmark in that exact spot matching the shape of a bullet wound. He has even found cases where the kid would have a second birthmark —the first matching a bullet entry wound and the second an exit wound. He has published volumes of work about this and they are seriously thorough, properly skeptical and truly fascinating.

Me: And what do you think?
Mom: I just don't know. It is interesting. And I was the most knowledgeable of the group about both cases. Everyone was confused about how I knew about this. I told them I have heard about nothing but Ian Stevenson and Jim Tucker for a year.
Me: Ha! And could any of you find "normal" explanations? Or any holes?
Mom: Both Ian Stevenson and Jim Tucker seem incredibly intelligent and did very thorough research. I just don't know anymore. It is kind of amazing.

And then I got another chance.

Growing up, I heard Gloria Steinem's name from my mom far more than my mom has ever heard about Jim Tucker from me. While I greatly admire her, my mom and the women of her generation admire her in a way I can't identify with. I grew up assuming, without question, that I was equal to men. My mom and her generation did not. My mom watched Gloria help change the world and would probably put her in her top five "want to meet" list.

So I was scrolling Twitter and came across, as the mediums, say, a gift from the universe.

Total "I told you so" gold.

A photo of Laura Lynne Jackson properly tagged standing with Gloria Steinem. Along with Jim Tucker and Ian Stevenson, my mom has been forced to become familiar with Laura Lynne as another household name. I took full advantage of the opportunity it presented; and I immediately sent a text with the photo of the tweet to my mom.

> ME
> Sooooo . . .

> MOM
> HOLY SHIT!

I called her. There was no way we were not going to have a conversation about this.

Me: Sooo, do you have anything to say?
Mom: I don't know. It's impressive. Laura just sent that to you?
Me: What? No. Why would she send that to me?
Mom: I thought Laura sent photos to you every day?
Me: Why would she send me photos? This was on her twitter.
Mom: But I thought Laura and Joe and all of the mediums sent you photos so you could twitter them out to other people?
Me: ???
Mom: Don't they twitter photos to you?
Me: MOM! Are you talking about the fact I do social media for Forever Family?
Mom: Yes. Isn't Twitter social media?
Me: Oh my fucking God! First of all you TWEET!! You don't Twitter. Secondly, you HAVE Twitter. You know how it works. No. No... they post their own photos on their own social media. And some-

times I share them, but I usually create unique content and… ugh… just never mind.
Mom: I'm glad you brought this up. I have been following what everyone is twittering but I can't figure out how to send a twitter myself. Can you show me?

A week later I was with my mom taking her through the most basic details of social media.

Mom: What do you have planned this weekend?
Me: Phran and Bob invited a few of us and some of the mediums to spend the weekend to hang out and go over some plans for the Forever Family Foundation.

My mom responded in a tone of voice usually reserved for speaking with children about the tooth fairy.

Mom: That is wonderful. Which ones are going to be there?

I replied with the enthusiasm of a kid talking about the tooth fairy, except after reading a book by a valid scientist who had actually seen the tooth fairy.

Me: Janet Mayer, Angelina Diana, Joanne Gerber, Laura Lynne Jackson and Kim Russo…
Mom: KIM RUSSO!?
Me: No… I cannot promise I will ask her. Just no.

My very skeptical, "only neuroscientists are on the correct path to identifying the true nature of consciousness" mother had developed, you could say, a soft spot for Kim Russo.

The reasoning behind this hard-won soft spot? Around October 2016, Kim Russo had posted an elaborate election prediction. It was not exactly specific and did not state who she was going to vote for, but it was clear the unexpected would happen and the world would be mesmerized. (Google it). Of all the crazy evidence I was gathering, this prediction was one of the least scientific things I shared with my mom. However, it became one of her favorites.

This was how our conversation went back in 2016 when I shared it with her.

> **Mom:** That's quite a statement. I guess if I thought there was any validity to this ridiculousness it would be a little scary. It sounds like Trump is gonna win.
> **Me:** Okay, we know Trump is NOT gonna win. We have seen the polls. Plus, he's batshit crazy. But there IS validity with Kim and all these mediums! Kim is one of the Windbridge mediums who passed Doctor Beischel's blinded tests. In fact, I'm doing my final paper for my class at The Rhine now on the quintuple blinded studies Doctor Beischel conducted. Wanna read it?
> **Mom:** Maybe later.

That meant no.

> **Mom:** So possibly there is some interesting thing Julie Beischel—
> **Me:** DOCTOR Beischel.
> **Mom:** Doctor Beischel has discovered, but self-proclaimed psychics have been claiming they can predict things for centuries.

FAST FORWARD TO NOVEMBER 9, 2016, when I was packing for the Forever Family Foundation Afterlife Explorers and Mediumship Convention in Florida.

In the middle of her rage and disappointment at the fact an orange madman had somehow been labeled president over a much more qualified woman, she put down the *New York Times* to look at me.

> **Mom:** So, umm, that medium? The one who made the prediction. Will she be there?

I looked up from my phone where my friends and I were sharing our devastation that said orange madman was labeled president over a much more qualified woman, exposing a level of sexism, racism, and who knows what else that we had no idea still existed.

Me: Kim Russo? No. She won't be there.
Mom: Too bad. Can you forward me what she said? And maybe ask the others what they think will happen?

FORWARD AGAIN TO September 2017 while I'm home hanging out with my mom.

Mom: I cannot believe it. Kim Russo was so on point.
Me: I'm impressed. You actually remembered her name this time.
Mom: I did. And I showed it to a few of my colleagues and explained one of YOUR psychics said this BEFORE the election.

BACK TO THE PRESENT, as I was leaving for Phran and Bob's.

Mom: Can't you just ask her what will happen? Will we be rid of him soon?
Me: It's not—I guess—respectful to bug them with these kinds of questions. I don't even know her. I dunno, if it comes up, I will ask. But more importantly, do you believe all this now?
Mom: I didn't say that. Stop exaggerating. Stop putting words in my mouth. I just want you to have fun. It's nice to see you not sad all the time.

BEFORE HEADING out I was careful to protect myself and shot off a quick email to Phran.

To: Phran
From: Liz
Subject: Can't Wait!!
Hey.
SO excited to see you! And thanks for including me. Quick reminder. I know I was able to be super open at the retreat about my loss but I still haven't had my reading with Laura and Angelina, and some of the others there, so I have to be secret about everything.

Phran replied.

To: Liz
From: Phran
Subject: Re: Can't Wait!!
Of course! Thank you for the reminder.

I knew I could trust her, but I liked to cover every angle.

When I got to Phran and Bob's, we sat around their table while everyone began to arrive. The conversation was lively and fun when Kim, who I was meeting for the first time, turned to me.

Kim: I'm curious, what's your sign?
Me: Scorpio. Oh shit. I uh—I should not have told you that!
Kim: Why shouldn't you tell me? What's wrong with being a Scorpio?

I noticed Bob looking very amused at how I was going to navigate this one. He was definitely enjoying watching me try to balance preserving my evidence, while not acting so weird I missed out on getting to know a cool new person.

Me: No, it's not that. I haven't had a reading with you, so I shouldn't tell you those kinds of things.

Kim: First of all, I'm not doing readings now. Second of all, what could I possibly know from your sign?
Me: I'm sensing something about the month of November. Is it a birthday?

Bob turned to Kim and pointed at me.

Bob: You don't know this one. She had to send in a paper check and fill out a piece of paper to book some medium sessions, and wore rubber gloves whenever she touched them.

This had become Bob's favorite story.

Kim: What!
Me: Fingerprints.
Janet Mayer: I ask for a paper check and piece of paper. Was one of those for me?
Me: Uh–oh–Yes. Yes, they were. But... ummm... what if you had an FBI connection or something?
Janet: I don't. How would I have this major connection?
Kim: Now I'm curious! Why does the FBI have your prints?
Me: They don't! As far as I know. But maybe they have a database on everyone's fingerprints. And maybe you have a friend or family member or someone who works for the FBI...
Janet: Don't you think the FBI has more important things to do with their time?
Me: Fine. Then the phone company. You use your thumbprint to open your phone and if you have a friend that works for Apple or Verizon and...

They were all staring at me baffled and amused.

Me: It's not like I know where your friends and family work!... OKAY! I don't think this anymore! At least not about you guys.

As we began to relax and have some wine, I turned to Joanne

Gerber, a medium I had had a reading with, but hadn't found a chance to get to know otherwise.

> **Me:** I still can't feel one hundred percent normal. Some days I am really happy again, which is something I never thought I would get to experience. And, then, other days I still miss my dad so much.
> **Joanne:** Oh, did he pass away?
> **Me:** Yes. I had a reading with you? One week ago. Do you remember he brought up my computer?

My dad did that every reading. At least every good one. He would bring up one small thing going on in my life. He had mentioned a dentist once by showing the medium a dentist chair when the day before I had been talking about how I needed to go to the dentist. With Joanne, my computer was getting repaired. Joanne had said that my dad wanted me to know my computer would be back in a few days.

> **Joanne:** I remember reading you. And I was surprised when you were saying what a skeptic you are because you were so open and easy to read. But I read so many people and I kind of go into another state. I can't remember the facts of anyone's reading.

None of them ever remembered readings clearly. I thought that was so interesting.

Phran left to go pick up Laura Lynne from the train station. Apparently, her car was being repaired. The rest of us continued to catch up.

I was living my sciencey/investigatey dream, getting to ask questions about how these otherwise normal people seemed to literally talk to dead people.

I had last seen Janet Mayer in Florida and it was great to hang out again after such an awkward start. I apparently had gotten so socially awkward trying to protect my evidence, she had thought I didn't like her! We both laughed about that at this point.

Janet also had an especially interesting story, which I had read about in her book *Spirits . . . They Are Present*.[1] One day, during a "Holotropic Breathwork" class, she had a bizarre experience. She started speaking an Amazonian tribal language. The words poured out of her, but she had no idea what she was saying. She ended up doing this regularly. After years of research, she found a person called Ipupiara (Ipu), an indigenous South American Shaman who held a Ph.D. in Anthropology and Biology. He was the first person to identify the tribal language. It was called Yanomami.

> **Me:** I wanted to ask you, do you understand this tribal language now?
> **Janet:** No. I don't understand any of it. If a group of people from that tribe were next to me talking, I wouldn't understand a word. Aside from the few phrases that I now have learned by normal means.
> **Me:** Seriously?! Wow. But when you are speaking, do you understand what you are saying?
> **Janet:** No.
> **Me:** But the words are pouring out of you fluently?
> **Janet:** Yes.
> **Me:** What in the fuck!? When you had your brain scanned by Doctor Tarrant, was either language part of your brain active?

Janet explained that apparently, the part of her brain that was active when she speaks Yanomami was the same part of her brain that was active when she gives medium readings. Neither the part where your primary language is stored nor where your secondary language is stored was active.

> **Me:** But how?
> **Janet:** I don't know. Weird, right?

As I was learning as much as I could about Janet's crazy experience, Phran walked in with Laura.

I have to say I was pretty proud of myself when it came to Laura.

She was the one I had known the longest (the first one I had reached out to for a reading—and the first one I had met in person) who also had the longest waitlist. I had done a good job keeping everything secret to preserve my reading, despite having attended three of her workshops and a few Forever Family Foundation events, even ones that had involved drinking.

I knew I could keep up these rules and skills with Laura tonight despite the fact we had started drinking and I was getting comfortable.

Angelina Diana, who I had briefly met in Florida but had not gotten to know, was here and I looked forward to getting to know her better. I turned to her.

> **Me:** Now, for real, not protecting me the way you might, if you were presenting at a grief retreat or anything. How do you know you are actually communicating with deceased individuals? What if there is some universal consciousness of love, and it presents itself in the form of our deceased individuals because that is how we understand love. But it isn't truly them and we don't actually continue as individuals. How would you know?
> **Angelina:** You mean like God?
> **Me:** NO. No! Nothing like God. There is no scientific evidence of a god. Nothing to do with anything that would require belief.
> **Angelina:** Well, I do believe in God. But, separate from that, I do honestly think it is individuals.
> **Me:** Because of belief or because of something more tangible?
> **Angelina:** It comes down to experience and what I feel. When I am communicating with those who have passed, it is a strong personality with the same feeling of differences in energy that I experience when I talk to people who are still living. They each have a unique and very real energy.

I asked Kim the same question.

> **Kim:** That would be deceptive for The Other Side to pretend to be your loved ones. The Other Side and the energies I communi-

cate with always feel so loving. Deception and trickery aren't loving.

Me: But, what if they are doing it to be loving? So we don't worry or feel sad. Maybe the only way they can express love to us is through imprints or images of those we love?

Kim: That's not genuine. Think about when things are false. You might feel excited about it, almost high, but then that goes away quickly, and you're left with an emptiness. Is that how your readings feel?

Me: No. Not the good ones. The good ones have a lasting depth. They add a kind of permanent healing that stays with me. The sadness doesn't disappear, but good readings add a feeling of happiness and love, as well as being intellectually fascinating. All of that stays and builds on top of this permanent sadness, and the happiness they help create becomes more powerful than the sadness, to where I can actually enjoy life again.

Kim: I think genuineness and honesty are required for something to have that deep and lasting effect.

We continued to drink and relax and slip into that buzzed silliness that is no different than any time I go out drinking with any other group of friends.

As they began to share embarrassing and funny stories that usually seem to come out when anyone has too much alcohol, they turned to include me. "Hey Liz. What about you? Any funny stories to share?"

Me: What? Uhhh—Oh right. Sure. Tons. But, first, I was thinking about the nature of consciousness and how it could survive outside a brain. I know I read somewhere they think the mechanism in our brains making this survival possible could turn out to be Microtubules. At least according to Sir Roger Penrose and Doctor Stuart Hameroff. And so, if you were to take your most evidential experiences and then...

Their faces said it all.

Me: Never mind.

I let my "scientific exploration" go, relaxed and joined in the conversation.

Kim started to tell me things about my life. I slipped into "sitter-mode" and tried to wipe away any reactions from my face.

Kim: I don't want to get your hopes up since this isn't a real reading. I would need to first meditate, then be alone with you and not have all this noise and energy around.

I did understand that and, considering the level of "poor conditions," the reading was impressive. After some more neutral topics, she went a little deeper.

Kim: I hope this is okay to say. I feel like you went through something hard that changed your life. I feel like you essentially had a breakdown.
Me: Yes.

I had just met her, but I felt totally comfortable with her saying that.

I also felt totally comfortable with the others hearing it, even though aside from Phran and Bob, I didn't know any of them that well. I was realizing that to do the kind of work they do, they had to be non-judgmental. I felt that there was nothing that a person had been through or no way that a person responded to a tragedy or trauma that made them uncomfortable.

Me: Yes. And that is what prompted me to explore and research this whole world. Are you getting what happened?
Kim: I'm not. I am sure if this was a real reading I would. But I'm not doing them now.

In my head, I screamed: SO MAKE AN EXCEPTION FOR ME! Out loud, I was more reserved.

Me: Well, uh, I guess if you ever do decide to do readings again or anything please let me know.

As we were getting a little drunker, I took the perfect opportunity to break everyone away from the fun again.

Me: What's the difference between PTSD panic attacks and knowing something like a premonition?
Laura: When I know something, no matter how bad it is, there's no emotion attached. It just is like, this is what happened or this is what's going to happen.
Me: Ah, okay. Thanks. It sucks because I keep expecting something to happen again and, every time someone doesn't answer their phone or is five minutes late, I panic.

While gathering logic like that does help, I was starting to accept that, unfortunately, there is no "cure" for serious trauma and grief. You just have to find a way to build your life around it.

Me: Thanks for explaining. And, seriously, thanks for bringing me into this whole world. It was pretty nice of you to do, considering when you first met me, I was such a mess. This is so embarrassing, but when I went to your first workshop, I had to be reminded to shower. I probably looked like I was a total mess. Or I guess at best like I was a lot of fun coming in after an amazing night out?
Laura: No not at all. You were actually well presented.

I WAS?!

We all stayed up late drinking and talking. I was grateful to have met some amazing people who I was now enjoying for themselves, not only as a connection to my dad.

THE FOLLOWING MORNING, I found Phran and Bob downstairs putting breakfast on the table.

Phran: Hey, can we pop into the other room to talk about something for a second.

I followed her into her office room.

Phran: I fucked up. Kind of big time.
Me: Tell me what's up and let's brainstorm how to solve it.

I assumed it was about some promotion for one of our events or, ugh, maybe she accidentally got us blocked on one of our social media accounts.

Phran: No, this is bad, and you will KILL me.
Me:...
Phran: Umm, you know how I picked up Laura separately yesterday? Well, umm, we started talking and then...

Laura was about to do an event for us with an investigative journalist named Leslie Kean next month. Maybe Phran had gotten overly excited and told Laura I was willing to do some massive promotion that would be a lot of extra work for me.

Me: Whatever it is we can correct things, or if you want me to do some extra work or something, I don't mind because...
Phran: Will you listen a second! You millennials are impossible to have a conversation with. You are here one minute and there the next. It is not just you. It's the same talking to my kids. I need you to listen.
Me: Okay! Okay!
Phran: Laura started telling me about her friend she's been trying to help. This friend of hers recently lost her dad and they were very close. She is about your age, so most of her friends don't get it and—
Me: Oh. Sure. I would be happy to talk to Laura's friend since she and I had the same loss and—

Phran: That's the problem I'm getting to. Will you listen! Stop interrupting me.
Me: Sorry. I'm listening.
Phran: So, I was like, "She should join this group called The Dinner Party." You know the one you told me you are a part of.

The Dinner Party is a support group for 20-and 30-somethings who have had a significant loss.

Phran: And then I said, "Liz also lost her dad and said The Dinner Party really helped her." And then Laura was like, "Oh no, Phran! I'm not supposed to know that! You HAVE to tell Liz immediately that you told me." Shit, I'm soooo sorry. It just came out.
Me: WHAT! OH! Ummm… OHHHHHHH FUCK!!… FUUUCK!
Phran: You are under a fake name right?
Me: Yes. But at this point, she will probably recognize my voice.
Phran: Maybe send a proxy or something?
Me: The proxy's deceased people could come in. And I want to experience my own reading.
Phran: Yeah. I know. I just feel so bad.

I knew how unintentional it was and how truly bad Phran felt.

Me: Don't worry. There's so much she doesn't know. I promise it's fine.

We headed into breakfast and joined everyone. Laura then pulled me aside.

Laura: I want to make sure you know that I know now what happened.
Me: Yeah, Phran just told me. It's okay.
Laura: I feel so bad. You worked so hard to not disclose anything to me so that we could have a clean reading. I know how much that meant to you.
Me: It's fine. Honestly, there's still so much you don't know.

Names. Personality. Memories. In a way, it will be a relief to not have to talk in code and circles anymore. Which I guess turned out to be pretty pointless.

So, to anyone sitting here desperately hoping this is all valid and trying to put this baffling puzzle together, here is another reminder that, while I cannot say I know where any of these abilities are coming from, you can eliminate fraud and dishonesty.

At least with this group.

28

When You Are Going Through Hard Times Just Keep Going

Joe Shiel, the medium/spirit artist I had met in Florida, was holding a one-day workshop on Platform Mediumship. Platform Mediumship is when mediums do readings for groups, and I wanted to learn more. It was being held at Forever Family Foundation medium Janet Nohavec's Spiritualists' National Union Church. Once you entered the small, nondescript stone building, it looked like a traditional chapel, a nod to Janet's early life as a Catholic nun, but with many different religions represented. It was packed with statues of angels and stained-glass windows, a Jewish star, and a Buddha statue symbol. Photographs of historic mediums, such as Edgar Casey, lined one wall.

Before welcoming the class, Joe gave me a hug and a warm hello. His love and reverence for the craft of mediumship was as clear in this classroom of students as it was to all of us grieving in Florida.

Joe S.: Were you sick? How was the weather? What happened that day? What was in the news? All this can affect a reading.

More questions immediately ran through my mind: Were the

readings affected because of how the medium feels personally about the news, weather and other everyday events? Do the energies of global events affect the world energy, as the Global Consciousness Project demonstrated, which then affects the connections between here and The Other Side?

> **Joe S.:** Everything you have gone through in this world, someone in the spirit world has also gone through. The more you have been through, the more you will understand what the spirits are sharing with you.

He continued.

> **Joe S.:** You need to have integrity to succeed as a medium: Be reverent and sincere. Treat the communication as sacred, so the spirits will want to communicate with you.
> **Me:** Why would it matter what the medium is like? Wouldn't my loved ones put up with an obnoxious medium to be able to talk with me?
> **Joe S.:** It's not that your loved ones wouldn't want to, it is that the energies do not connect strongly, even if they are trying. I can't say exactly why.

I guess it was kind of like an old cell phone that can't get a good connection. Or maybe the medium might miss details of the subtle communication from The Other Side, the way arrogant people often miss the cues and needs of others on this side.

When we broke for lunch, I went to talk to Joe. Since he often spoke about ethics in mediumship, I wanted his opinion.

> **Me:** I went to Lily Dale for a séance this summer and it was clearly bullshit. I wondered if you had any insights or anything.
> **Joe S.:** I've never heard good things about the séances there. Who did you see?

I told him.

Joe S.: That guy is a total fraud. He has even been arrested for being a fraud before.
Me: That's what I thought. But do you think he could have real abilities too?
Joe S.: No. And even if he can have real abilities, who cares? What he does is incredibly unethical. He is taking people's money. He lies and he tricks them.
Me: But what about the fact it makes them feel better?
Joe S.: Lots of scams can make people "feel" better. A fake doctor or fake investment pyramid scheme both make people feel good for a little while. He is taking people's money under false pretenses. That's fraud.

After lunch, Joe continued sharing his ethical values and respect for mediumship.

Joe S.: We need to remember that the spirit world has the big picture, which we do not see here. I read for the family of a murder victim, but then I also saw the murderer and saw the murderer as a human, as a baby and as someone's baby. This is not to excuse murder in any way, but the spirit world sees the full complexity and layers of all situations. Everyone has a story and a humanity.

That was a hard one. It is definitely easier to hate people who do evil things than see the tragedy of their whole lives and recognize their humanity. Probably the crueler someone is, the more tragic their life has been.

I got insight into something I had been worrying about for some time. During a few readings, when the medium had been highly accurate, they would suddenly throw in a generality that cold readers often add, like I had an upcoming trip.

This was the reason: There was a famous medium who had mentored a lot of the now big mediums. In fact, the medium in that movie Poltergeist was based on her, and she would teach mediums to have "fillers," to avoid silence. Mundane things like "you should get your car looked at" or "you have an upcoming trip" would be thrown in while waiting to get the next piece of genuine information.

Joe did not think these fillers were a good idea. Silence was preferable. I agreed.

When Joe directed us into our first set of exercises, I joined three middle-age women. If gave my usual spiel about being sciencey and not having abilities. I started to say what came to mind, but I didn't have that wavy feeling. I felt as if I were just guessing.

Me: I'm getting a grandmother and I see her making cookies.

I was not trying to make them think I had abilities, but I gave an unintentional cold reading. I said the first thing that came into my mind and it was coming from the logical part of my brain that would reasonably deduce that someone of a certain age had lost a grandmother who baked. While she said I was right, and continued to say so about more information I got, it all came from logical deduction, and I made sure to be clear to her that's all it was. I think that explains some unintentional fake psychics who are unaware of how good they are at deducing this logical information.

Joe concluded the class, leaving us with the humbling sacredness and huge responsibility of being a medium.

Because my bus wasn't scheduled to leave for some time, I sat in the hall and started to read. I was engrossed in Dr. Claude Swanson's book *The Synchronized Universe: New Science of the Paranormal*. Dr. Swanson was a physicist educated at MIT and Princeton. Along with his career in applied physics, he set out to pursue the mysteries of the paranormal and afterlife. He explained his Synchronized Universe theory using the analogy of a fan. Imagine you didn't know what a fan was. When the blades were spinning, you wouldn't see them as individual blades. You would not even realize the blades existed. Now, if you were in a dark room and a strobe light was flashing in time to match the spinning blades, you would think the blades were standing still. In either scenario, you wouldn't know the actual reality of what was actually going on.

Joe noticed me and asked me to join him. I followed him into a small and cozy office and sank into the couch, he took a seat across from me.

> **Joe S.:** So, what did you think of the class?
> **Me:** Amazing, inspiring, fascinating. I honestly am still a bit baffled by this whole consciousness out of a brain thing.
> **Joe S.:** I saw that you got accurate information today. That's what those women in your group said.
> **Me:** No. No, I didn't. I said stuff they agreed with, like that they had lost a grandmother. Obvious stuff that came from a logical part of my brain.
> **Joe S.:** You should pursue this. I think you could be a great medium.
> **Me:** A few of you guys in Forever Family have said that I had abilities. I always have taken it as you guys just being kind and welcoming, but I don't think that is what you mean. I do not have abilities, aside from a few anomalies, and I am still not sure I even believe all of this.
> **Joe S.:** I'm not flattering you when I say that. You remind me of my son, who is a bio-engineer. He doesn't believe any of this because he cannot see how it works. But he's brilliant. And I see you have that kind of intelligence. You think deeply about everything and try to look for their honest explanations. That's what the heart of being a medium is about. And, if you listen and focus your intelligence and thoughts that way, I think you could make real connections.

I was touched and humbled by his words.

> **Joe S.:** Mediums are like lights. And I see you are like a light. You have that *thing*. But what does a light do?
> **Me:** Illuminates?
> **Joe S.:** It illuminates, but also exposes. You have the benefits of illuminating, but also the difficulties of exposing. It's just who you are. You expose people who are full of it without even meaning to

and that makes false people uncomfortable. People's ugliness gets exposed. Just like a medium.

Since I was little, I had problems with certain social scenes or people, even when I was trying my best not to. My school, from kindergarten through seventh grade, was a chilly place filled with conformist values, intense New York City social climbers, and a lot of cruelty. I did have some great friends and had known some wonderful people, but at the same time, I seemed to offend certain teachers, kids, and their families, even when I had tried my best to placate.

It took me until after college to realize it was when I had ideas and opinions I was confident expressing, seemed to not care what people thought, or when I was excited and energetic about anything new and creative—anything out of the grey or power hungry norms of that society—that this group would get their sensibilities so deeply offended.

Me: Thank you. Seriously. I've also been wanting to ask you a question: how do you know that all of this comes from individual consciousnesses? How do you know our consciousness doesn't blend or fade, or that it is not just some universal love that we interpret as those individuals we love?

I was looking forward to the reassurance of Joe's answer, because along with being a skilled medium, he also questioned everything, deeply and thoughtfully. Like me, and the rest of the people going to see him, he had also suffered a loss: his soulmate had recently died.

Joe S.: Honestly? I don't know. I genuinely hope so, especially after the loss of Robin. I don't know why she hasn't talked to me yet either, since she was a medium herself. I assumed as two mediums we would be communicating regularly. But I'm asking everyday what this is really about and I'm more motivated than ever to get answers and to grow my abilities and become a better medium.

And there it was. The trade-off for my need for the truth and my need to explore this deeply. And my trade-off for developing friendships with the mediums. That was not the comforting reassuring answer. That was the honest answer.

He also shared some remarkable experiences he had had.

About 30-years ago, when he was starting out in New York City living with his then-girlfriend, he used to take the subway to work. His route went under the Twin Towers. He would get anxious visions of them collapsing on him, but he never got specific dates or other revealing details. He was affected enough by these visions that he ended up taking different routes.

> **Joe S.:** Those panics were awful. Also because of having been in war, I have PTSD.
> **Me:** I have PTSD too.
> **Joe S.:** I'm sorry to hear that. You know how it works? The trauma makes a black dot, which shows up in brain scans. Then when something triggers it, that black dot gets activated and spreads like waves over your whole brain. That is why PTSD feels as if it's taking over the way it does.
> **Me:** That's exactly what it feels like, but I have never heard about that black dot. I just assumed I have PTSD from what happened, and the fact I now have panic attacks.
> **Joe S.:** If you don't mind my asking, what happened?

I paused. I had had a reading. He had gotten so much about the core of my dad's personality and the energy of our relationship, but he had not gotten our actual relationship. I deliberated for a second, but then decided, "fuck it." I wanted to go further with what was an important conversation.

> **Me:** I know the relationship did not come up in our reading, but I lost my dad. I have been dealing with intense PTSD since losing

him. He went pretty quickly too. Not in an instant, but he got sick and then, a few weeks later, that was it. I can barely breathe anytime my mom doesn't answer her phone, or if she has a doctor appointment. I keep waiting for the next shoe to drop. It does feel exactly like this wave of panic that takes over my brain and then my body. It's awful.

Joe S.: Grief is just the worst. I'm going through it, too. I'm not sure you have actual PTSD though. This is just part of grief and, you know what, it takes a really long time and is very hard. Excruciating actually. It's that line—when you are going through hard times, just keep going. That's what you and I both have to do.

29

The Afterlife Luncheon

I took the train out to Long Island for the Forever Family Foundation Afterlife Luncheon. Laura Lynne was giving readings and Leslie Kean, an investigative journalist and *New York Times* bestselling author, was giving a talk on her recent book, S*urviving Death: A Journalist Investigates Evidence for an Afterlife.*[1]

As I helped guests check in, a woman I recognized from Florida gave me a big hug.

Woman: How are the medium readings going?
Me: Fascinating! And healing.
Woman: So, you're taking clients?
Me: What?
Woman: Your readings. Are you taking your own clients now?
Me: Wait, no. I meant I have been GETTING readings. I don't give them.
Woman: Oh! I remember that reading you did in Florida. You were so startled, but so happy when you gave it. I thought you would have been pursuing your own mediumship after that.
Me: Oh God. NO!

This is something I have come to feel strongly about. Giving a good reading once, or even more than once, does NOT mean you should try to provide evidence to grieving and desperate people that their loved ones are still around. That could be devastating to someone in grief. Yes, I did get inexplicable information more than once. But I can't get the level the mediums get and I don't get it consistently. Unless someone can frequently get verified, evidential information, understand when they get psychic instead of mediumship information, can relate to what mediums mean when they say they have had to learn to "turn it off," they should not even attempt beyond classroom exploration. (None of this applies to me.)

Leslie took to the stage for her talk. She was refined and intellectual in a non-pretentious way. She looked exactly how you imagine a grounded journalist would—short hair, glasses, a blazer, and a calm and thoughtful poise. She spoke about everything from Jim Tucker and her experiences with physical mediumship to medium readings that shocked her, including one reading she had with Laura and a second she had with a medium named Sandra O'Hara. I held a soft spot for Sandra, after she had evidentially brought in my childhood cat (early on when I was still super skeptical). She had described his fur color, his flat, very bossy face, and haughty, entitled personality.

Overall, Leslie was blown away by how, once you start putting all the evidence together, there really seemed to be an afterlife!

During the question-and-answer session, I raised my hand.

Me: Maybe I'm projecting from my own experiences, but I went in assuming that none of this was true. Then I had different levels of first—What the fuck—oh sorry for saying fuck—umm I mean, wow, oh my god, moments of complete bafflement. It was like maybe there *is* something going on worth pursuing. Then I went further and there was the next level of tipping point to okay something REALLY is going on here. Have you had that and, if so, what was that tipping point?

Leslie contemplated for a moment.

Leslie: I would say that Laura and Sandra, two mediums who don't know each other, getting the same information about my brother was pretty impactful, but it really comes down to putting all of the evidence together. I can't think of one specific point or moment, but it built up bit by bit as I kept accumulating more and more inexplicable experiences.
Me: Oh, and one more question, about physical mediumship. The one you thought was genuine—Stewart Alexander. Was he in a cabinet?

(The cabinet was the makeshift tent that Séance Medium sat in at the Lily Dale séance.)

Leslie: No. He sat with us in a circle.

There was a quick break before Laura's reading.

Another volunteer and I both took a mic to opposite sides of the banquet hall, where we waited to bring the mic over to whoever Laura felt guided to read.

Laura's readings were as remarkably evidential as ever, and my impatience to get a reading with her was as strong as ever, but that stomach-dropping shock I always got just wasn't there. I felt an intellectual fascination and I felt the emotions of the guests' losses, but those 'HOW IN THE FUCK!?" chills all over my body were gone. I tried to will them to come back. But they didn't.

After Laura finished her last reading, she started answering questions.

One woman was barely able to speak through her tears. Laura waited, patiently and kindly.

Woman: I lost my son last year...

My mind was screaming, *UGH! DON'T TELL HER!* But some people are looking for emotional comfort more than strong evidence. Actually, we were ALL looking for emotional comfort, but some of us need a higher level of evidence to get it, while some of us need more nurturing.

> **Woman:** Mediumship and signs are so important to me, but nobody around me believes any of this.
> **Laura:** And, so, you feel really alone with this? And need someone to talk it over with?
> **Woman:** Yes.

I could certainly identify.

> **Laura:** But you have your friend, Amy. I'm getting that name. That is who you talk about this with right?

Laura had never met this woman before.

> **Woman:** Yes!!
> **Laura:** The Other Side is bringing you and Amy closer. One of the things they do is help bring us the support systems we need to manage our grief.

I looked behind me at Phran and Bob, who were watching to make sure things ran smoothly. I knew I could talk over my experiences, doubts, and questions with them anytime. Melissa, another volunteer, smiled at me as I looked back at her.

And watching Laura do the scientifically impossible, the thought that fraud was behind it was no longer a consideration. She might know about my dad, but I also knew there was zero chance she was Googling me. I thought of all of the other mediums I had gotten to know over the past year, and I could say the same thing about them. Maybe a certain magic and awe of early discovery was gone, but it was replaced by a wonderful support system, one that would never have occurred to my dad to guide me toward while he was living on this side.

I intercepted Leslie as she headed over to join Laura at the book signing table.

Me: Can I ask you a not-in-front-of-everyone question?
Leslie: Sure.
Me: This physical mediumship stuff. You saw it. And you honestly do believe it happened? Because I saw some of it and it seemed… well, like a total bunch of crap.
Leslie: Most of it is. Who did you see?

I told her.

Leslie: That one you saw is a total fraud. He has even been caught before.
Me: Yes. I heard that.
Leslie: Was it in total darkness the whole time?
Me: Pitch black. And he was hidden in a curtained cabinet.
Leslie: Most are fake, but that one I saw, Stewart Alexander, really blew me away.

At that moment, Phran came up to us.
"Liz—come on! Leslie has to go sign books NOW."

Me: B… b… but we are talking about physical mediumship.
Phran: I need you to go sit with her and help oversee the signing.

She told me that another volunteer was sitting with Laura.

Me: So, I get to sit with Leslie?
Phran: Yes. And thanks, Liz. But remember she and Laura need to sign books and talk to guests. Not answer your questions.

In between signings, I continued to ask Leslie questions.

Me: So, I have another not-in-front-of-everyone question. What do you think? Super-psi? Or do you think our individual consciousness actually continues?
Leslie: I guess this is something we can never have proof of, but I do think it is individual survival.

Me: Why?
Leslie: Survival is the only one of those two that can explain drop-ins. You know when a medium is giving a reading and a person they don't recognize comes in and they later find out many given facts matched a specific person. Like the Emil Jensen case.

During the event, Leslie had talked about the Emil Jensen case, a fascinating story she wrote about in her book. The Jensen case took place in 1905, when a highly investigated Icelandic medium named Indridi Indridason was holding a group seance where he received information from a man who had passed away and was now, as the mediums say, "in spirit." Indridi Indridason shared the information he was "receiving," from this deceased man, but no one in the group recognized this person as one of their loved ones. This deceased being gave the name Jensen and said he had been a manufacturer. He then described a fire in a factory in Copenhagen, which was brought under control an hour later. A few weeks after the séance, newspapers arrived by ship, detailing the fire in Copenhagen, which had—just as the medium said—broken out in a factory and was contained within an hour. The date of the séance was November 24, the same day the newspaper reported that the fire occurred. Emil Jensen (the one who had passed away) apparently "gave" many personal details to the medium as well. About 100 years later, an author found the careful notes kept from Indridason's séance and matched the public records stating details of Emil Jensen—name, age of death, living siblings, the fact he was a manufacturer, and even his address, which was on the same block a fire had broken out on November 24th.

I had read about other mind-blowing drop-in cases as well.

Me: But that still could be super-psi or a general consciousness of love.
Leslie: True. But, from everything I've studied, I really do think it is survival.

I guess we all have to make peace with the fact we can never have 100 percent proof.

30

The Right Seat on the Right Flight

"Well Fuck!"

That was the title of my email to Phran after I got an alert that my flight to the Grief Retreat in Florida was canceled the day before I was supposed to fly out due to a snowstorm.

I tried to find flights for the next day. None.

And none for the following day.

Fuck! Fuck! Fuck!

I called the airline. Since everyone else was calling, I couldn't get through. I decided to escape into a movie and try again after it was over. I found a cute indie film with an actress who looked like an actress I was friendly with in LA. I looked at the names to see if it was her, but it wasn't. The actress' name was Diana. Not my friends' name, but it was my grandma's name. I then noticed the actress looked a lot like Laura Lynne; the character was an English teacher, just as Laura had been before she pursued mediumship. I reminisced about the time Laura had inadvertently possibly given me a message from my grandma, which she had no idea she had even done. That memory and the fact that this actress shared the same name as my grandma got stuck in my head and replayed in one of those annoying loops. I mean, I'm always thrilled to "hear" from my grandma, but after a brief thought of "love" to her, I just wanted to escape into my movie!

The character, who was about my age, was dealing with the loss of her mom. I thought about how hard that must have been for my mom to have lost her own mom when she was still a teenager. When a song played over the closing credits, I became absorbed in its 1970s pop sound. I realized that my roommates had played this song just the other day. For some reason I had listened intently, zoning out and absorbing myself into the lyrics as my mind slipped into that weird meditative state it goes into whenever I would get signs. I rarely do that with music unless a song stirs up special memories, but this song was irrelevant to my life.

My mind did the same thing now, as the song played along with the credits.

I dreamily thought about this song, how my roommates had just played it, how Laura had possibly brought in my grandma randomly during a conversation we once had, and the fact that this actress shared a name with my grandma, which was now rolling across my screen.

A final screen popped up—a dedication screen—"For Anna."

My Mom's name! Was my Grandma saying hi to her?

I knew my Mom would dismiss this as a bunch of foolishness, so there was no point in telling her.

But, at that moment, I really wanted to make it to Florida. I tried the airline again, and I sent a request out to my Grandma and any connection I could think of on The Other Side for help.

After more than an hour on hold, a woman finally helped me change my flight. I was supposed to leave the next day (Friday,) but my early morning flight was canceled, and the afternoon flights were filled. The best she could do was to book me on a flight to Florida on Saturday morning.

We hung up. I opened my email—and seriously!? She had me on the new Saturday morning flight, but returning to New York on Tuesday, a day earlier than I had planned. I normally double check everything and would have asked her to stay on while I opened the email, but I was so tired.

I called back and waited. And waited.

This time, it took an hour and forty-five minutes to get someone on the phone. I explained my situation.

Airline Rep: We actually DO have a flight tomorrow, Friday. It's leaving at ten am. We had to cancel all earlier flights because we need the morning to clear the ice off the runways. There's one seat left. Do you want it?

That seat must have just opened up since that option was not available before.

Also weird: I had apparently misread the email. The first representative I spoke with had not moved me onto an earlier return flight. In fact, she hadn't changed my flight at all. I was still booked on the canceled flight. I'm not sure how I misread the dates. Organizing dates, travel, paying attention to details is something I have always been especially good at.

The only issue with this new flight was that the one available seat was a middle.

Fine. I could certainly deal with that.

Mom: So, what happened?

I told her.

Mom: You know that is weird. Maybe there is something to all your stuff.

I was taken aback by her reaction. There was no pattern to when she said something was me being delusional and something was worth considering.

Me: I asked Dad and Grandma for help, even the Auxiliary Board.

The Auxiliary Board is made up of the deceased loved ones of

Forever Family board members and deceased scientists who help Forever Family Foundation from The Other Side.

> **Mom:** Okay, okay. Enough. I can't go that far into all of this. But I don't know… things like this do make me think.

The next morning, I left for the airport super early. I went to check in at a self-serve kiosk, but for some reason, it showed an error. I went to check in with an airline official. Since I was there anyway, I figured I would take a shot in the dark.

> **Me:** I know I got the last seat on a packed flight, but there aren't any window or aisle seats available? Are there?
> **Gate Agent:** Yes. We have one aisle if you don't mind the back.

Had this aisle just opened up? If there was anything to this, my loved ones must have been working hard for me.

I made it onto the plane and ended up seated next to a cool-looking middle-aged woman with tattoos and an edgy sense of style. She was flying with her teenage daughter.

I put on my headphones and popped an anti-anxiety pill and closed my eyes, hoping for a smooth flight and, if I couldn't have a smooth flight, could I at least have a flight where I wouldn't die. I am not the best flier.

The plane took off. With the recent storms, it was one of the roughest take-offs I had ever experienced. My hands were shaking and I was trying to breathe deeply. The woman next to me tapped my arm. I took off my headphones.

> **Woman:** Excuse me. I saw that you looked terrified. It's just the winds from the storm. But you can hold my hand if you want to.
> **Me:** Oh… Uh, yes. I actually would like that. If you don't mind. Thank you.
> **Woman:** This is the first flight I've taken without my husband in over twenty years. And it's my daughter's first flight ever without

her dad. It's actually the first flight I have taken since he died six months ago.
Me: OHHHH… I'm sorry to hear that. In that case, let me tell you where I'm going.

And I did. Carefully. So I didn't sound like some person selling the "answer."

Woman: I do believe in things like signs, which he has sent me. And I had a visitation. It felt so real. I'm a nurse, so I have seen people have some inexplicable deathbed visions. But, because I'm a nurse, I'm pretty scientific and I cannot understand how consciousness could function without a brain.
Me: There are these scientists Sir Roger Penrose and Doctor Stuart Hameroff, and it's very early-stage research, but there seems to be something in our brains called microtubules that could, I guess, download? consciousness from another source. And there's this woman, Doctor Julie Beischel…

And I filled her in on all I had been learning. She listened and asked questions. Her eyes took on a hopeful glow.

Maybe it was not for me at all that The Other Side was making sure I got just the right seat on just the right flight.

31

In Spite of Stephen Hawking

I finally made it to Florida, and I joined a table of guests just in time for dinner.

As usual, the guests were primarily middle-aged and older women who have lost a child or husband, and some couples who had lost a child. A few were joined by a young adult child, grieving a father or sibling.

The conversation turned to the difficulties we've had with the way society deals with grief.

> **Me:** How about the "I-don't-know-what-to-say" face? As if everyone doesn't die. EVERYONE. So why is it so shocking and uncomfortable to others when someone does?
> **Woman 1:** How about this one—I'm a mom to four. Three who are still here and one who passed. I have had people say, "At least you have your other kids." As if that makes it okay that I have had one die?
> **Man:** How about, "You should be over it by now. It's been a year."

Bob then stood up. He introduced himself and Phran, and the three mediums—Joe Shiel, Janet Nohavec, and Doreen Molloy.

Bob: And how many of you thought there was NO chance of life after death.

I raised my hand and looked around.

Me: Wait, am I the only one?
Young Woman: No. Back here. Me too.

The voice belonged to one of the few other young women. She was there with her mom. Maybe also grieving her dad?

Me: So just the two of us.
Bob: So, Liz, what do you think now?

This was the big question: when it all came together, what DID I think? I paused. I was constantly re-evaluating what I thought.

Me: At this point, I think that there is a preponderance of evidence pointing towards survival and, when I put it all together, it's a very strong possibility. Actually, no. While I can't say it's a definite, it is a probability. A strong and very realistic probability.

Before dessert, I went to go meet this other non-believer.

Me: Hi. I'm Liz. So it looks like it's just us!
Young Woman: Apparently. I hope you are right. I lost my dad and I miss him so much.
Me: I did too! It's the worst.
Young woman: I'm twenty-eight and I feel like I'm an adult. It shouldn't be this hard. But, I dunno, losing him I feel like I need him as much as I did when I was three. Is that weird?
Me: I am so happy you said that because I relate. I saw a girl about eight years old walking with her dad holding hands and I... I almost lost it.

We hugged, then I returned to my table for dessert and to continue talking with my tablemates.

> **Woman 1:** Last year I had the best reading at the Forever Family Foundation event. My daughter loved the holidays and loved Passover. She had made this little blue bowl and we had used it for parsley, the bitter herb, at our last Passover seder together. He kept describing this blue bowl and saying he saw this green leaf spice or herb in it.
> **Me:** Who was the medium?
> **Woman 1:** I can't remember his name. He's a young guy. From New York.
> **Me:** Joe Perreta! He is amazing. Super evidential. And he's not even Jewish.

During my reading, Joe was the first medium who got details of my childhood housekeeper/nanny. She had recently passed away, and he was the first reading I had had after she passed. He then brought in my dad who made us laugh a bit as his personality came through.

> **Joe P.:** Your dad… he's funny. He's reminding you that when you get mad, or have to deal with difficult people—you had to deal with a lot of that when you worked in fashion—not to get mad or anything. That you learned from him what to do.

He let Joe know he would say something witty with instigatey people, which had the whole room laughing at them. Or with them, when the person was innocently difficult. My dad had a brilliant knack for that kind of wit and timing.

> **Woman 2:** I've had two wonderful readings, but I still wish Stephen Hawking would come out and say something about all of this afterlife research.

This was before Hawking died in March 2018.

Of course, I wanted that too. But, at this point, how much would that change things for me? Yes, it would still make a difference, a big difference. But not even the opinion of someone like Stephen Hawking could change everything I had seen and experienced.

32

Coin Collections and Connections

I grabbed my seat in the circle of folded chairs and settled in for Joe Shiel's workshop.

> **Joe S.:** Think of what the color red means to you.

I thought of a red dress I own that I was wearing when I met one of my ex-boyfriends. Which also made me think of him.

> **Joe S.:** Now, each of you thought of something different, which is why universal signs do not work and also why mediums can misinterpret signs your loved ones are using to communicate.

He shared heartbreaking stories about the level of tragedy he has worked with up close: families who had lost all three of their children, people who had lost their entire families.
And he shared a personal story of hope.
We all knew that he had lost his soulmate, Robin Bennet, recently and that he was right there grieving with all of us.
Joe loved to sail, but like all of us in grief, he had been uninterested in his usual hobbies. He hadn't sailed at all in the six months

since Robin's passing. One day, he was driving with Loyd Auerbach, when another driver sped up, jumped lanes, and cut him off like a maniac. Understandably, Joe was shaken. He sped up to the car to tell off the driver. That's when he noticed the car's license plate, which read, "Sail JS."

Sail, Joe Shiel.

Needless to say, he was no longer angry.

He then dove into readings where, among others, he brought in an ex-boyfriend the woman had not talked to in a decade.

As we filed out, I overheard this woman's conversation. "I have not thought about him in years," she told her friend. "Joe was definitely not reading my mind."

We then headed into Janet Nohavec's workshop and readings.

Janet N.: And here is one of the truths that is not easy to accept: Just because you are a good person does not mean bad things won't happen to you.

That touched a nerve. One that was intensified every time I heard of some horrible person living their lives. Every time I saw Ivanka happily standing with her dad, The Orange One. Harvey Weinstein. Brock Turner. One thing that no one, from mediums who seemed to defy the laws of the universe, to even Einstein himself, could begin to touch upon with any validity was why some people die young when others don't.

Then Janet dove into readings.

When she got to me, she brought in someone who became clear was my dad, mentioning among other things, his amazing sense of humor and that he saw me surrounded by papers and "research." I might not have it in paper form, but yes. And she described my dog, Peanut. Even though Peanut was still living, it made sense that my dad would bring her up.

Earlier this year, Peanut had gotten very sick. The vet said I could try some medical treatments, but they did not recommend that path, since they were unlikely to work. Still, I did the medical treatments. That morning, when I came home from dropping

COIN COLLECTIONS AND CONNECTIONS

Peanut off in the hospital, I saw my dad's name written in chalk on the subway steps! No one ever writes anything on these subway steps. At the end of that day, his name was still clearly there, as if it were freshly written, even though that subway stop is a high foot-traffic area. Then, despite the minuscule odds, the shot-in-the-dark treatments worked, and Peanut got better.

> **Janet N.:** Do you collect stuff, or did your dad used to collect stuff? Coins or something?
> **Me:** No?
> **Janet N.:** Why would he be giving me silver dollars or half dollars?
> **Me:** Coins? Is he talking about coins?
> **Janet N.:** Yes.
> **Me:** Umm, I'm investing in Bitcoin and other crypto lately.

I had also been consulting with a startup that was looking to work with Blockchain in fun and creative ways to crowdsource data.

> **Janet N.:** You are?! Get out of here! That WOULD be why he's talking about coins.

She seemed to enjoy the novelty of this meaning of 'coin.'

> **Janet N.:** Your dad wants you to know he is around. That is why he keeps bringing up all this stuff. He knows how much you question it.

She did not say that to anyone else. He still knew I wanted my evidence.

33

The Best Medicine

Later that evening, after a walk around an indoor labyrinth and a sound bath meditation, we headed in for wine, cheese and a chance to mingle. The wine mixed with the intense release of so many emotions all day, which created a bubbly and lively type of buzz.

One woman told us about her hilarious son, who apparently had come through in all the readings today and made the entire group laugh. She reveled in getting to enjoy his personality again and was delighted that the other guests had an opportunity to enjoy him too.

The woman I had spoken with earlier joined in—the one with three children still living, the fourth on The Other Side.

> **Woman 1:** So, do you wanna hear the most awkward moment I ever had with my kids?

Yes. We wanted to hear.

> **Woman 1:** I took all four of my kids on a road trip to Disney World. On the way, we found this motel, not expensive, but it seemed perfectly fine. I notice all these women in little skimpy outfits and huge fake boobs. I come to find out that they had a

porno convention there. The younger ones were really young, and we were on our way to the hot tub, when I see what's going on. I notice there are two women and a man having a threesome. Then another woman pops up from under the water, obviously giving the man a blowjob underwater! My young kids couldn't understand why I suddenly wouldn't let them play in the hot tub.

Needless to say, we were all laughing.
Bob decided to stick to the theme of the weekend.

Bob: Couldn't that woman have drowned? How would she explain that one to a medium?

At that exact moment, Joe Shiel came down.

The mediums are all unique, wonderful and wise people. Each one that I have gotten to know holds a place in my heart in a different way. That being said, there are the ones who will go out drinking and will get really silly, and there are those who do not.

Joe is not an uptight person by any means, and he has a good sense of humor, but he is not one of the go out drinking, get silly ones. He actually never drinks. In an ideal world, there were a few other mediums much better suited to answer this question. Rebecca would have been my first choice.

But Joe was there.

Me: Joe. Hey Joe! Hey, I have an important question.
Joe S.: Okay. Yes, Liz.

Maybe at the end of an exhausting day, he might not have been up for a "Liz question," but he didn't show it.

Me: Let's say you are doing a group reading. And someone's loved one comes in and you found out they drowned in a hot tub by giving a blowjob. Would you say it?
Joe S.: Seriously, Liz! Seriously? I don't know.

I have had that exact response to my questions before, but never with that intonation of a mix of slight amusement and dismissal at the silliness that Joe had now. My questions usually force mediums to delve into physics and the meaning of the universe, which this one certainly did not.

Despite the lack of "scientific depth" of my question, I have to admit, that woman impressed me. She is in year one of her loss and she was okay laughing and even had all of us laughing. I hardly laughed at all my first year. And when I did, I felt so guilty immediately afterward that I stopped quickly. But, once I could laugh again and I realized it wasn't wrong, I felt so much better, at least for a little bit.

34

Still Getting Enlightened on the Other Side

Doreen Molloy hosted the final workshop of the weekend. After sharing a bit about mediumship and her abilities, she dove into readings.

Since I had already had a private reading with her, I was curious to see how this reading differed from my first. I knew she consciously did not remember me but were parts of my reading stored in her unconscious?

Because I had had readings with so many Forever Family Foundation mediums and now so many of them knew what happened, I also wondered if that was affecting my readings? I now can say that I know these mediums were not intentionally sharing info on me, but what if communities had some consciousness bank that they tapped into. That would explain things like groupthink and why sometimes people, who were otherwise kind, could get swept up into group values and commit horrible acts of cruelty.

During our earlier private session, Doreen had brought in my dad. Today, she brought in my grandmother, my mom's mom. After describing identifying characteristics, such as the fact I look like her and that I wore a ring of hers as a necklace, she got further into the messages my grandmother wanted to share.

Doreen: She is very proud of you and what you are doing. The work you are doing. She is saying, "If I was still here I would NEVER do something like that." Let's just say she is enlightened now and very aware of what you are doing, so keep doing that."

Doreen did not know me, my skepticism, or what my grandma could be talking about.

After Doreen's session, I headed into the closing ceremony, where guests had a chance to ask questions of all the mediums. The questions asked were typical of the ones everyone asks when they first learn about all of this. The same ones I had asked.

Guest 1: How can you tell if you are actually reading someone who passed away or if you are just reading the sitter psychically?
Doreen: I get asked this all the time and it's a great question… and this is one of the reasons why I agreed to be a 'human test subject' with the Windbridge Institute and before that, with the Veritas Research Program at the University of Arizona, Tucson. I participated in many double-blind and triple-blind studies over a period of years. During these types of laboratory controlled blinded studies, the medium being tested has no knowledge of the deceased, nor do they have any knowledge of who the sitter is… and yet accurate information still comes through during the testing protocol. Even more importantly, there is often information that comes through [during these blinded studies] that the sitter is unable to verify—until they're able to speak with other people who might know whether certain information is in fact 'accurate or not'… and in most cases, they are later able to confirm that the medium was indeed accurate. So to answer your question, I can't be 'reading the mind of the sitter' if the sitter had no knowledge of that information to begin with. This indicates that the information is clearly coming from 'somewhere else.'

I turned to the woman next to me who had asked the question.

Me: There are also the studies of Doctor Jeff Tarrant, where medium readings showed up in one area of the brain and psychic in another. Check them out!

Guest 2: What happens if your loved one has reincarnated? Can they still come in during a reading?

The mediums again gave various answers, as best they could. Essentially, the answer is that time and space are an illusion, so even if someone has reincarnated it would not matter and they could still communicate.

Doreen gave an answer similar to the one Laura had given me early in my research describing a maypole, except using the metaphor of a bicycle wheel. Each spoke was a life and the center of the wheel is the core of who you are.

Bob: Any more questions?

Me: So, I have learned a lot about the survival of consciousness shortly after death, but what about long term? Like in five-trillion-years. Do we still maintain our individual memories? And when did our consciousness first originate? Were we conscious before the Big Bang and, if so, where did we live? And when the sun burns out and our planet is no longer habitable does our consciousness go exist in other material solar systems or as a non-material entity? Or what about after the Big Crunch? Wait, why are you guys all exchanging looks and laughing?!

Doreen: If I could answer any of these, I would have a trillion dollars. Do you guys wanna take this one?

Joe S.: I don't know.

Janet N.: I don't know.

Well shit! I had hit a wall. My questions were no longer answerable.

After the Q&A, it was time for one of my favorite parts of the weekend, but also one of the hardest. The memorial video that I had made with photos of the guests' deceased loved ones.

I had sat with every single photo, looking at each one and wondering about who they were, what did they like to do? Who were their friends and what inside jokes did they have with them? I would look at their faces, were they laughing and vibrant? Or did they look more contemplative? It was such a different experience to see the same photos at the end of the weekend. As I watched the video in the room, a certain abstraction was ripped away. I had now spent an entire weekend bonding with the living loved ones of the people in these photos and hearing the stories of their loved ones. The reality of these profound losses overwhelmed me, but it was soothed with more than just wishing that these people would be reunited.

At the end of the day, that was what this quest for scientific evidence of an afterlife was really about—these people, these mothers and fathers, sisters and brothers, husbands and wives, and children, who were the whole world to somebody in this room.

35

It's Not an Addiction. It's Science.

I spent that night at Phran and Bob's home in Florida, where we unwound from the intense weekend.

> **Phran:** By the way, Liz, how many mediums have you gone to at this point?
> **Me:** Not that many actually. It's not like I go to one every day or anything.
> **Phran:** Who would go every day?
> **Me:** Right. That's what I said. I don't go every day.
> **Phran:** But that shouldn't even be a consideration.
> **Me:** I know. That's why I said I don't go every day. Meaning I don't. Not that I do.
> **Phran:** How many?
> **Me:** Hmm… Lemme see… Not too many. Thirty-eight. At this point thirty-eight.
> **Phran:** What! Nobody should go to thirty-eight mediums. Especially not in one year because—
> **Me:** Year-and-a-half.
> **Phran:** Will you stop interrupting me. You should learn to make your own connection. That is what Forever Family Foundation is

about teaching people to do. Go to a good and genuine medium to get evidence that this is true, then make your own connections.
Me: But it's—
Phran: I know you are about to say it's science.
Me: I was. It IS science. I need to explore and research thoroughly. These people...
Phran: I think you are addicted. You are a psychic junkie.
Me: I AM NOT! I go a lot because it's not only shocking to have normal people defy the laws of the universe right in front of my face, but I also like them as people and I want to be able to get to know them without having to be all secretive.
Phran: But, if you didn't need to keep getting readings, you wouldn't have to keep things secret.
Me: I talked to my dad every day, so once a month is not that much.

Both of us had put down our wine glasses. Too immersed in the conversation to keep sipping.

Phran: That is what I'm saying. You can talk to him without going to a medium. You also are getting amazing signs. You have gotten your evidence and learned to connect. Don't you feel happy about that? Do you still really want to keep getting readings?
Me: No! I don't want to keep getting readings. I want to be able to call him and hang out with him, not through mediums and not through signs but as himself in his own body.
Phran: Well, of course. Everyone would prefer that.

Of course, that is what everyone wants, but nobody gets that. Unless you die young yourself, everyone eventually has someone die that they would give up every other part of their life to have not die. Yes, I can be sad, and a part of me always will be. But, if I let myself become bitter like that, it will ruin my life. I realized that there was no point going down that road tonight or ever again. I know that is not possible all the time, but I can at least hold it as a personal standard.

IT'S NOT AN ADDICTION. IT'S SCIENCE.

Phran didn't need to tell me that. Just watching the way she and Bob live their lives is a constant reminder and inspiration to not waste my life feeling sorry for myself.

Me: In all honesty, I do love the emotional connection during a reading. And, because each medium that I meet is so unique, each reading ends up being special and different than any other reading. While I don't get shocked the way I used to, it is still astounding, even if in a different way now. Anyway, scientists and researchers are supposed to collect a wide range of data.
Phran: Well, if you love the science so much, why don't you start reaching out to the scientists?
Me: Because it hasn't been an option. It's not like I can go onto their website and book a session to accompany them on data research.
Phran: Okay. That is a point. But you would, if you could?
Me: YES! Oh my god. I would LOVE to! I wanna…
Phran: So maybe those are the things you need to start doing now. The questions you're asking at the retreat of the mediums, they can't answer them. You should talk to Claude Swanson.
Me: For real!? I would LOVE to talk to Claude Swanson.
Phran: Okay. I will see if I can introduce you. Let's get to bed. We are gonna wake up early and Donna will be here at nine. We need to start planning next year's Grief Retreats and Mediumship Convention. It will be a busy day.
Me: Goodnight.

I got into bed, fell asleep and slept deeply through the night.

Afterword: WTF Do I Even Think?!

My cousin and I were sitting in the kitchen, catching up a few weeks after I got home from Florida.

> **Cousin:** So, I am curious because I do trust your opinion and it is all pretty remarkable what you have seen. What do you finally think? Is there actually an afterlife?
>
> **Me:** Honestly, and you won't love this. I don't know. Not with one-hundred percent certainty. But, I think there is, without a doubt, a different set of laws of the universe than we think or, I guess, perceive as humans.
>
> **Cousin:** Like what?
>
> **Me:** We very probably survive death, but I can't say definitely. If I could guess, our individual consciousness is downloaded by our human body. It's stored in some bank of consciousness somewhere, where we probably have experiences and relationships, and those people that we connect with probably come from the same section of the bank. Our brain downloads this conscious-

ness the way our nose, for example, "downloads" scents. A scent is not created by our nose; it perceives the scent from a source.

Cousin: This is really interesting!

Me: Right! And there are probably infinite states of being that this non-local, non-physical consciousness is "downloaded" into for "us" to have an experience. Sometimes to humans on earth, sometimes to beings in other dimensions, sometimes to material states of existence and sometimes to non-material states. Sometimes, the material planets are right here in our own galaxy and sometimes they are in other solar systems or galaxies or universes in an infinite multiverse. I think this is the best way to describe it for now, but it's probably impossible to clearly describe and understand while we are living in this dimension of consciousness any more than I could describe a new color.

Cousin: I always wonder if we all see colors the same or if my red is your blue, or maybe we all see completely different colors. For example, my blue is a color that doesn't even exist for you.

Me: I know! I wonder the same. But then with all of this, when I take everything I have read and seen and experienced, and each person I have met, and I put it all together, it seems impossible to say that the materialist definition of the world is even close to all there is. And so many others who have researched this draw the same conclusion for the same reason. Plus, now I know I can eliminate fraud or delusion from the people I have met. Once I knew I could trust these people, the body of evidence became massive. And there is still so much more to explore.

Cousin: I am so glad you are doing all the work. It is really interesting. But I would never actually take the time to explore myself.

Me: I really love it! I wanna keep going. Like think about this—I just had a talk with Phran. She said that we are way too focused

Afterword: WTF Do I Even Think?!

on the brain. How do we even know that is where consciousness originates, or where the real ability of mediums takes place? What if it's actually all hidden in some single cell in our stomach or something.

Cousin: Let me ask you what you ask everyone else. What is the strongest piece of evidence you have that makes you think the afterlife is a high probability?

Me: Every single thing is like a piece of a puzzle. Some pieces are larger than others. To get this whole picture, you have to put them all together. The medium readings and their personalities, the signs I have gotten and that others have gotten, the years and years of tests going back to the work of the Society for Psychical Research in the eighteen hundreds to the work of Doctor Julie Beischel today. Reports of Near Death Experiences and Out of Body Experiences. Past life memories and the research of Doctor Jim Tucker and Doctor Ian Stevenson. All I learned at the Rhine and the years of psi studies and the Global Consciousness Project. The amount of anecdotal evidence, too, from random threads on Quora to stories from friends of friends to unknown blogs. They all have parallels. In fact, I heard Doctor Beischel say in a podcast once that a bunch of anecdotal evidence is still evidence.

Cousin: Yeah. I see what you mean about the puzzle.

Me: There is this weird consistency to all of it too. The feeling of energy that people across the board describe, or this random person will post to a forum about their kid having past life memories and it matches what Jim Tucker found, such as the age of the kid, the emotions attached to the memories or the ordinariness of them. The physical sensations, such as the waves I have experienced, are consistent with what others have experienced, even before I knew that was what it was "supposed" to feel like.

Afterword: WTF Do I Even Think?!

I got up from the kitchen table and refilled my glass of water, before joining her back at the table.

Me: Also, there's this beautiful simplicity and mundaneness to all of this. It is not this fantastical stuff like memories of being a queen or the absurdity of that Lily Dale séance. It's like these shocking things happen, but with this subtle and delicate tone. This quantity of evidence and a consistency among it seems to be why most people conclude this is real.

Cousin: Another one of your questions back at you. What was the number one turning point for you?

Me: Florida, when I gave that reading. It was not just the reading in and of itself, although it was obviously mind-boggling, but the way it transformed my understanding, which allowed me to take in the information I had been gathering at a deeper level. Things that seemed too foreign to begin to understand started to make sense. And things I had blown off as probable bullshit were suddenly more relatable, such as when mediums or researchers or OBEers (Out of Body)ers explain their experiences or describe that wavy, buzzy, hot energy I felt, or when they say that truly explaining all of this is impossible with the limited language we have for these kinds of experiences.

Cousin: Oh, yeah, that Florida thing was so batshit! And what about Stephen Hawking and the best neuroscientists? If this absolutely paradigm-shifting stuff had any validity, wouldn't they be all over it?

Me: Ya know, that one still gets to me. And I don't think there is a black-and-white answer. In all honesty, I think most have already decided it's bullshit and they have never bothered to examine the evidence. I would love to know what they would conclude, if they dug as deeply as I have. But a big part of that would have to be getting to know the people. It's not only the things that happen,

Afterword: WTF Do I Even Think?!

but the people behind them. When you know them personally, you cannot dismiss what they say.

Talking with my cousin, and everyone else who prefers science over faith (myself included), I know one issue for accepting this evidence is that a lot of this cannot be repeated consistently in labs or follow the actual standards of science. For example, I cannot repeat what happened to me in Florida or Renee's class. The mediums can, but not perfectly. They will all get some things wrong or have "off days." Even the best ones. But it still bothers me that they don't study these mediums. And I still do worry about what they would conclude.

There's also a ton of snake-oil and believeyness in this world. Science needs to be protected from that and from religion. And scientists don't want to lose their grants.

But I think it comes down to the fact they are often not that interested and therefore, have never bothered to dig deeply into this.

I realized that I never had an actual opinion on any of this afterlife stuff before. I thought I did, but it all really came down to "it is not true, because Stephen Hawking and friends say it is not true." Not because I ever had researched myself. And that means what I thought was my opinion was just another form of faith and belief.

Me: I still feel as if I'm at the very start of my research though.

Cousin: I can't wait to see what you learn as you continue. By the way, do you wanna come with me to this party tonight?

Me: YES! I just have to finish a few things first. I have to get this Instagram strategy plan over to Phran and Leigh and a proposal for our 2018 Grief Retreat.

Cousin: I love seeing you enthusiastic again. This has transformed your life. And brought you back to life.

Me: It has. I feel as if I'm a different person living in a different

Afterword: WTF Do I Even Think?!

universe, but I have no idea what that looks like yet. I think that's one of the biggest misconceptions about healing grief. There is this idea that healing means getting back to your old life and who you were before your loss. But, for me, my old life doesn't work anymore. I think healing has been more about transforming into this new person with new values, but oddly I feel like I'm more myself than ever before.

Cousin: I can see that.

Me: I just need to figure this out and what my purpose for being here is. I used to think it was about having as much fun as possible and succeeding at a fun career, making a ton of money, and having fun with guys, and, of course, one day finding the right one. But I feel as if it's so much more than that now. I definitely wanna meet the right guy and have a family and a career that brings me joy and money, but I also want real meaning. Not only do I feel like a different person because of my dad, and that the laws of the universe are so different than I had thought, but I also have met so many parents who have lost kids who would be about our age. I feel there should be some purpose or meaning or… I guess… it's as if I have some responsibility since I'm here and they're not.

Me: I also keep thinking of what Christine said to me early on. That I know I feel like everything is over and done, but think of it like a movie, which opens with a huge dramatic thing that the character thinks means everything is over, but really it's the start of an incredible journey or transformative experience. She was so right! While my dad dying was and is the worst thing that ever happened to me, learning and exploring all this afterlife evidence is one of the best things that has ever happened in my life.

Me: Before, I could never have dreamt in my wildest dreams that we probably survive death in some way, and I cannot think of anything I would want more than that.

Afterword: WTF Do I Even Think?!

Cousin: Yeah? Honestly, I am glad you're happy again.

Me: Thanks! But I would give it all up to have my dad never die in the first place.

Cousin: Of course. That goes without saying. But that is probably why humans don't get to make those kinds of choices. If we did, we would never accomplish—what is it the mediums call it? A higher calling? Or a soul plan?

Me: Did you just agree with mediums? Did you just use the word soul?

Cousin: Fuck off.

Teaser: WTF Just Happened?! Book 2

WTF Just Happened?!:
A sciencey skeptic investigates even more evidence of an afterlife and fights for justice with her dead dog.

That evening after dinner, I walked out of the banquet hall and joined a line of people who were waiting to walk the labyrinth.

At all our Florida Retreats, we transform one of the conference rooms into a labyrinth. This involves unfolding a massive cloth labyrinth. Following the spiral-shaped path puts your brain waves into a meditative state.

Only a few people are allowed in at a time. The lights are dimmed, relaxing music plays, and everyone holds electric votive candles with "flame" bulbs because fire is not allowed in the hotel. Melissa shakes wind chimes as each person enters the labyrinth, slowly walking, in their own world, processing the intensity of the weekend, and maybe? connecting with their loved ones.

I always go once the room is empty, so I can record videos to post on our social media. I stood quietly by Phran as the guests came into the room, a few at a time, and lost themselves in their own world. Or, more likely, a world with their loved ones.

When the room was empty, aside from me and Phran, I stepped into the labyrinth. Despite having my phone on, recording the pattern of the walk, I slipped into a meditative state, just as I did every other time I had walked through the labyrinth. I tried to connect with my dad. They say when the brain is in a meditative state is when it's easiest for the discarnate consciousness of our loved ones to communicate with us.

I gave my dad the usual assignment that he still hasn't "accomplished."

Me: Dad! Tell me something mom knows and I don't, that I can verify with her.

As I began to walk, I started to "feel" my dad. I smelled him and saw him in my mind's eye, with his huge warm smile. I felt that sense of vibrancy and safety I often felt around him. But I can conjure that up anytime. That is not evidence that my dad is actually still conscious and communicating with me.

Me: Dad! Come on!

Then I stopped seeing? imagining? him and suddenly my little crazy dog, Peanut, who was a pug chihuahua, AKA a chug, popped into my mind. As soon as I saw her little eager face with her wide eyes and tongue sticking out, she began running in circles around my feet, getting "the zoomies" all around the labyrinth. Then I "saw" her run back to me, eagerly dancing around and looking at me, expectantly, the way she always would when she wanted treats, approval, head scratches and being asked her (and probably most dogs) favorite question, "Who's a good girl?"

I didn't *see* her see her, as if she was alive, but I very clearly imagined her, in a way that I couldn't control. As if she had pushed herself into my mind, and I had no choice but to have the visions I was having. I also felt deeply connected to her and felt that light giggly silliness, combined with that overwhelming feeling that is best described as "just not even being able to handle how cute she is" that I always used to feel with her. This "experience" of Peanut stayed with me until I circled back to the exit of the labyrinth. And then there I was. Back in the conference room. The feeling and images of Peanut were gone.

Me: Phran! OMG! It was so cute. I kept picturing little Peanut running around me the whole time. I couldn't stop laughing.

Phran had always loved Peanut. Even if she hadn't met her in person. Or in "dog," I should say. She always asked how she was doing, wanted to see photos, and wanted Peanut to say hi on Zoom. Or, I guess pre-2020, on Skype.

Phran: WAIT!!!! Did you just say Peanut??
Me: Yes. Why?
Phran: Holy shit! I kept seeing Peanut with you. I thought as you started, "Oh, of course, Liz will connect with her dad," but I looked and there was Peanut!
Me: Are you kidding! No way!

Phran was not even a medium and she saw Peanut??

But while Phran was not a medium, she had always "known" things. She once "knew" that she and Bob were going to win a car in a raffle. And, tragically, the day Bailey passed, she woke Bob up at 4 in the morning in terror, gray, and shaking, stating "something terrible is going to happen today." Which not only makes one question the possibility of everyone having "psychic abilities," but also question the nature of time. If we can know the future. And to what extent we can alter the future. Or if certain things are just "written."

We stared at each other in amazement. My mind began to race.

Me: But Phran... what do you think? Do you think Peanut was really here? Which is great evidence for the survival hypothesis. Or were you reading my mind?

Phran started laughing.

Teaser: WTF Just Happened?! Book 2

Phran: You and these fucking questions. Can't you ever take the evidence? Or the experience?

Follow me as I continue to investigate evidence of an afterlife in the most awkward and skeptical ways possible:

Order *WTF Just Happened?!: A sciencey skeptic investigates even more evidence of an afterlife and fights for justice with her dead dog.*

www.wtfjusthappened.net

Science + Spirituality Salons

Wanna meet me (either virtually or in person), learn more about the science and get a reading with a highly evidential medium?

Host a private Science and Spirituality dinner or brunch in your home or virtually.

This is an intimate, highly emotional, and evidential experience. It consists of Liz sharing her experiences examining evidence of an afterlife and a scientifically certified psychic medium giving group readings.

Learn more here:

Science + Spirituality Salons

Or email me:
    liz@wtfjusthappened.net to learn more.
    I hope to meet you!

Stay in Touch!!

If you have a second would you please rate and review this book on Amazon, GoodReads, Apple, etc. It makes such a huge difference in the algorithm!

I cannot tell you how much I appreciate this.

If you want to hear conversations with some of these amazing people who changed my mind about an afterlife, and get to hear me bombard them with questions (as you can tell, that's my dream life!) check out my podcast on all podcast apps.

WTF Just Happened?!: All about the afterlife. No woo.

Join me at a Science + Spirituality Salon, and/or reach out to me with any questions, comments, to just say hi!
    I love hearing from all of you. liz@wtfjusthappened.net

Stay in Touch!!

Notes

1. How Hard Can Time Travel Really Be?

1. Amber Stuver, "Einstein's twin paradox explained," TedEd, September 26th, 2019.
2. Wikipedia. "Bell's Theorem." 22 November 2021. https://en.wikipedia.org/wiki/Bell%27s_theorem
3. Flatland: The Movie. Directed by Dano Johnson and Jeffrey Travis. 2007.
4. "Searching for the Science behind Reincarnation." NPR, NPR, 5 Jan. 2014, https://www.npr.org/2014/01/05/259886077/searching-for-science-behind-reincarnation.
5. Jim B. Tucker, M.D., *Life Before Life: Children's Memories of Previous Lives* (New York: St. Martin's Griffin, 2008).

2. I'm Not Done Talking About Reincarnation

1. Wake Up. Directed by Chloe Crespi and Jonas Elrod. 2010.
2. Nelson, R. D (2017). 'Princeton Engineering Anomalies Research (PEAR)'. *Psi Encyclopedia*. London: The Society for Psychical Research. https://psi-encyclopedia.spr.ac.uk/articles/princeton-engineering-anomalies-research-pear. Retrieved 26 November 2021.
3. Roger Nelson, "The Global Consciousness Project Meaningful Correlations in Random Data," The Global Consciousness Project, https://noosphere.princeton.edu/.

4. A Few More Books Get Me Thinking This Isn't All Batshit

1. George Musser, *Spooky Action at a Distance: The Phenomenon That Reimagines Space and Time —and What It Means for Black Holes, the Big Bang, and Theories of Everything* (New York: Scientific American and New York: Farrar, Straus and Giroux, 2015), Kindle.374
2. Diane Hennacy Powell, M.D., *The ESP Enigma: The Scientific Case for Psychic Phenomena* (London: Walker Books, 2009).
3. Ibid., 228.
4. Dr. Lisa Randall, *Warped Passages: Unraveling the Mysteries of the Universe's Hidden Dimensions* (New York: Harper Collins e-books, 2009).
5. George Musser, *The Complete Idiot's Guide to String Theory: Take Your Understanding of Physics into a Whole New Dimension!* (Pennsylvania: Alpha, 2008).
6. Dr. Robert P. Lanza, *Rethinking Immortality* (The World and I Online, 2013).

7. 15. Ray Bradbury, *Dandelion Wine* (New York: Bantam Books, Reissue edition 1985).
8. Dr. Robert P. Lanza, *Rethinking Immortality* (The World and I Online, 2013), location 42, Kindle.

5. The Sacred Scriptures Of Gary And Julie

1. Gary E. Schwartz, Ph.D, *The Afterlife Experiments: Breakthrough Scientific Evidence of Life After Death* (New York: Atria Books, 2002). | Paul Davids, Gary E. Schwartz, Ph.D., *An Atheist in Heaven: The Ultimate Evidence for Life After Death?* (Los Angeles: Yellow Hat Productions, Inc., 2016).
2. Julie Beischel, Ph.D., *Among Mediums: A Scientist's Quest for Answers* (Arizona: Windbridge Institute, LLC. 2013).
3. Ibid, Chapter 1: On the Same Page. Location 131.
4. Laura Lynne Jackson, *The Light Between Us: Stories from Heaven. Lessons for the Living.* (New York: The Dial Press, 2015)eu.

6. I Go To My First Medium And Wtf?!

1. Dr. Dean Radin, *Entangled Minds: Extrasensory Experiences in a Quantum Reality* (New York: Pocket Books, 2009).
2. Ibid., 230.

7. Did I Get A Sign?

1. Paul Davids and Gary E. Schwartz, Ph.D., *An Atheist in Heaven: The Ultimate Evidence for Life After Death?* (Los Angeles: Yellow Hat Productions, Inc., 2016), Chapter 7 Sean Asks for a Sign.
2. Bernard D. Beitman M.D. author of Connecting with Coincidence, "Can You Accurately Estimate Coincidence Probabilities," Psychology Today, April 2nd, 2019. ; Shankar Vedantam, "Magic, Or Math? The *Notes* 375 Appeal Of Coincidences, and The Reality," Hidden Brain Podcast, NPR. 2017 https://www.npr.org/ 2017/05/08/527442620/magic-or-math-the-appeal-of-coincidences-and-the-reality.

8. Searching For Psychics And Ghosts

1. Fiona MacDonald, "There's Evidence Humans Didn't Actually See Blue Until Modern Times," Science Alert, April 7th, 2018. https://www.sciencealert.com/humans-didn-t-even-see-the-colour-blue-until-modern-times-evidence-suggests.
2. Claude Swanson, Ph.D., *The Synchronized Universe: New Science of the Paranormal.* (Poseidia Press, 2003).
3. J. Richard Gott, *Time Travel in Einstein's Universe: The Physical Possibilities of Travel Through Time* (Boston, MA: Mariner Books, 2002).

Notes

9. Weirdness At A Medium Workshop

1. Sarah Pruitt, "The CIA Recruited 'Mind Readers' to Spy on the Soviets in the 1970s," History.com, October 17th, 2018, https://www.history.com/news/cia-esp-espionage-soviet-union- cold-war.
2. Stephen Hawking, "Stephen Hawking Discusses Shadow People," May 26, 2014. https://www. youtube.com/watch?app=desktop&v=Xpjg6Aheqfg.
3. Wikipedia. "Religious and philosophical views of Albert Einstein." October 27th, 2021. https://en.wikipedia.org/wiki/Religious_and_philosophical_views_of_Albert_Einstein#
4. Wikipedia. "Michele Besso." November 27th 2021, https://en.wikipedia.org/wiki/Michele_Besso.

11. Spoons, Psychics And WTF?!

1. Loyd Auerbach, *Mind Over Matter: A Comprehensive Guide to Discovering Your Psychic Powers*. (Woodbury, MN: Llewellyn Publications, 2017).
2. 33. Ibid., 265.
3. J.B. Rhine and J.G. Pratt, "A Review of the Pearce-Pratt Distance Series of ESP Tests". https:// ia800307.us.archive.org/20/items/NotesonSpiritualismandPsychicalResearch/AReviewOfThePearce- prattDistanceSeriesOfEspTests.pdf.

12. Five-Dollar Readings And Three-Hundred Dollar Candles

1. "James Randi on Astrology." June, 16, 2006. https://www.youtube.com/watch?v=3Dp2Zqk8vHw. 376

14. Energy = Mindblown Squared

1. Charles T. Tart, "Laboratory Studies," in *Psychic Exploration: Out of Body Experiences* (New York: Cosimo Books, 2016).
2. Robert A. Monroe, *Journeys Out of the Body* (Norwall, MA: Anchor Press, 1977).

20. More Mediumship And Past Life Regressions

1. Janet Nohavec, Suzanne Giesemann, *Where Two Worlds Meet* (Chula Vista, CA: Aventine Press, 2011).

26. Smoke and Mirrors and Séances

1. https://nsac.org/what-we-believe/principles/.

27. Told-You-So's and Secrets Revealed

1. Janet Mayer, *Spirits . . . They Are Present*. (Bloomington, IN: AuthorHouse, 2011).

29. The Afterlife Luncheon

1. Leslie Kean, *Surviving Death: A Journalist Investigates Evidence for an Afterlife*. (New York, NY: Crown, 2017).

Also by Elizabeth Entin

Book 1 (This Book)

WTF Just Happened?!: A sciencey skeptic explores grief, healing, and evidence of an afterlife.

Book 2

WTF Just Happened?!: A sciencey skeptic investigates even more evidence of an afterlife and fights for justice with her dead dog.

Podcast:

WTF Just Happened?!: All About the Afterlife. No Woo.

Acknowledgments

Thank you for making sure this book was actually readable and helping me sort out the next steps to getting it published— **Miles Doyle, Renee Buck, David Tabatsky, Jen Houle, Jenna Gundersen, Gemma Leghorn, Kim O'Hara, Arestia Rosenberg, Michelle Morrow**

Thank you **Amanda Guerassio**, of Studio Guerassio, for being able to so perfectly capture this brand visually.

Dad: This book is for you. What a fun adventure we are going on together.

Mom: Thanks for not trying to put me in a mental hospital and finally considering I might be onto something.

Phran: More than I can put into words, love and misses, my dear friend and mentor.

Bob: The one person who was as skeptical as me in your early days. Seeing another logical-minded, left-brainer believe all this helped me so much.

To all #FFFCertifiedMediums: Thank you for putting up with me and my endless questions and skepticism. Thank you for being honest, genuine, and normal enough that I believed you. You are changing the world for skeptical people like me more than you realize.

Joe Perreta: Thanks for reminding me I am not constantly hallucinating, and that, yes, this really happened. Thanks for doing so many worldview-changing experiments with me and listening as I endlessly analyze theories of consciousness (even at 4 a.m.)

Renee Buck: Thanks for putting up with me in your class while I asked endless questions and always disrupted everyone during the readings with my level of shock. Also, thanks for being there to mutually vent through the four years of orange tyranny. Our texts, calls, and long hikes helped keep me sane.

Rebecca Anne LoCicero: Thanks for always making me laugh, for putting up with me and for making sure I always have my organic food whenever I stay with you. And Julia! I'm so glad your mom introduced us.

Janet Mayer: Thanks for still liking me after I was so weird and awkward when we first met. Thanks for giving me such a good reading, even when you couldn't have access to my fingerprints.

Joe Shiel: Thank you for your wisdom, ethics, encouragement and patience with my endless questions and skepticism.

Laura Lynne Jackson: Thank you for always being so nice to me in my skeptical days and for not throwing me out of your classes when I kept interrupting with tons of skeptical questions. And of course, deepest gratitude (more than words) for bringing Phran into my life.

Gina Simone: Thank you for your patience with me, my skepticism and atheism in your workshops and for answering all (well at least some:) of my questions.

Angelina Diana: Thank you for evidential readings at the retreats even though you didn't wanna have to read the "sciencey-one," and for putting up with me in your workshops.

Dusten Lyvers: Thank you for helping me know it was okay to "come out" early on.

Kim Russo, Joanne Gerber, Doreen Molloy, and Janet Nohavec: Thank you for being such evidential mediums and for taking such a grounded approach to all of this. You have helped change my world view.

To the FFF volunteers and board members: Leigh Harris, Annette Marinaccio (my fellow left brainer), Johanna Suchow, Donna Mello, Dr. Mo Hannah, Tom and Melissa Gould, and Robin Murray (I miss you!): What a great group of people to explore alternate dimensions with.

To Skye and Amy: While in one sense I wish we never had to meet, I can't think of anyone I would rather be with in the "dead parent club."

Loyd Auerbach and John Kruth: Thanks for teaching all this "paranormal" research in such an accessible, logical, and scientific way.

Huge thank you to Dr. Julie Beischel and Mark Boccuzzi for creating The Windbridge Institute and The Windbridge Research Center. Without the data and studies, I would never have given any of this a chance. Thank you **Dan Sturges** for an insightful conversation so early on in my grief. Thank you **Leslie Kean** for dedicating your logical, investigative journalist mind and skills to this phenomena, which all too often gets dismissed.

Harper Spero: thank you for making me realize it would be much worse to have murdered someone than to be researching weird "woo" shit.

Thank you to everyone in Harper's Circle with me—**Lillian, Michelle, Gemma, Heidi, Amy, Mallory, Kat, Gaby, and**

Mary: thanks for offering great advice as this book came together and for keeping me sane during my social isolation until I was vaccinated.

Thanks to the **many researchers** over the years, who took this crazy shit seriously enough to investigate it in a science-minded way.

Thanks to **Dreamers and Doers** for being an amazing resource for all my entrepreneurial endeavors, and making new projects less daunting.

To my writing group—**Allison, Dani, Jessica:** Thanks for helping me make sure this book didn't suck, for having people to write with and for endlessly fascinating conversations.

To all my friends (you know who you are): thanks for believing me and not thinking I have completely lost my mind. (Or at least still hanging out with me even if you do think I have).

To anyone reading this book: thanks for giving this world a chance. I did, and it changed my life more than I can explain. Don't lose hope.

About the Author

Liz is a writer, podcaster and entrepreneur who splits her time between New York and Los Angeles. She runs the social media for Forever Family Foundation and continues to research the evidence of an afterlife.

To follow her story, listen to the podcast, book a Science + Spirituality Salon, and stay updated:
www.wtfjusthappened.net

- facebook.com/WTFJustHappened6
- instagram.com/wtf_just_happened_
- youtube.com/@wtfjusthappened
- tiktok.com/@wtf_just_happened_0
- amazon.com/stores/Elizabeth-Entin/author/B0BGSWN7TP?

www.ingramcontent.com/pod-product-compliance
Lightning Source LLC
Chambersburg PA
CBHW031313160426
43196CB00007B/506